THEMATIC CARTOGRAPHY AND REMOTE SENSING

THEMATIC CARTOGRAPHY AND REMOTE SENSING

Edited by
PRITHVISH NAG
Deputy Director
National Atlas and Thematic Mapping Organisation
Calcutta

CONCEPT PUBLISHING COMPANY, NEW DELHI 110059

ISBN 81-7022-410-1
First Published 1992
Reprinted 2002

Published and Printed by
Ashok Kumar Mittal
Concept Publishing Company
A/15-16, Commercial Block, Mohan Garden
NEW DELHI-110059 (India)

Lasertypeset by
Microtech Advance Printing Systems (Pvt.) Ltd.
H-13, Bali Nagar
NEW DELHI-110015

Dr. G.K. DUTT

PREFACE

Thematic cartography has taken a place of a movement in India. Thematic maps and atlases are new being considered as an end product of major academic, scientific and planning activities. Individuals and institutions who were marginally interested in this applied field are now taking much interest. National Atlas and Thematic Mapping Organisation (NATMO) obviously became the hub of such activities. Under the guidance of its founder director, Professor S.P. Chatterjee, this institution fought tooth and nail to make a place for thematic cartography in the Indian scientific community.

Over the years, NATMO became a model institution in the world. It is the only institution which has a mandate to prepare the national atlas in particular and thematic maps in general on a continuous basis. Since it is in a developing country, the responsibility is greater. Its activities are watched, assessed and become models for the thematic cartographers in other parts of the world.

The survival of thematic cartography is not only linked with the development in the related disciplines, but also due to the technological impact it is facing or likely to face. The specialization of the thematic contents in maps and technological options have proved to be complimentary to each other. The objective of the present volume is to consider the issues and challenges involved and to discuss the technological options and solutions. It also takes into consideration the growth of the subject and also the scenario in the near future. The issues have been related to the modern technology, including mainframe and personal computers, compact disks, remote sensing, digital mapping and combination of technologies, such as the geographical information system. This volume will clearly give a message that the progress in thematic cartography will have to be linked with the modern technology and remote sensing, hence the name of the volume. The name is also linked with Dr. G.K. Dutt, ex-Director of NATMO. He had been associated with this institution from its formative stage. His role in promoting the thematic cartography has been exceptional. He has been exponent of thematic cartography with technological advancement. The testimony of his contribution is the monograph on the *Scenario of the National Atlas & Thematic Mapping Organisation in 2001 A.D.* I had the privilege of being its co-author. The present book has rightly been considered as his felicitation volume – a token love, respect and authority he commands.

In order to felicitate Dr. Dutt, I have been supported by my friends and colleagues who either know him personally or through his contributions. I got favourable response overwhelmingly. The authors readily accepted my invitations to write on a suggested topics, though they had to make time out of their busy schedules. The authors are recognized for their contributions, and belong to well known research institutions, such as, Calcutta University, DST, ISRO, IIRS, NATMO, NRSA and the National Library. Three authors are from abroad – Australia, Kenya and the United Kingdom. I am highly obliged to them. I am particularly grateful to the National Remote Sensing Agency and the Indian Space Research Organisation for allowing me to reproduce their imageries and other informations in the form of jacket and annexures.

This work could not have been completed without the support from all corners. My colleagues in the Ministry of Science and Technology, particularly Shri Y.S. Rajan, Advisor, Shri S. Ghose, Joint Secretary, Shri M.S. Sokhanda, Director and Dr. R.K. Midha, Director, were a constant source of encouragement. My friends in NATMO, such as Shri R.K. Banerjee, Dr. N.D. Bhattacharya and Shri G.N. Saha always stood beside me in my initiatives. I am grateful to them. Further, I was benefitted by the co-operation from the members of the NATMO Library. My close association with ICA, IGU, ITC, ICSU, ICSM, ISPRS, London School of Oriental and African Studies, University of Edinburgh, University of Durham, Space-Time Research (Melbourne), International Journal on Geographical Information System, and the like have been extremely advantageous. Through these contacts I could keep in touch with the technological innovations in thematic cartography.

7 August 1990 PRITHVISH NAG

CONTENTS

LIST OF APPENDICES

LIST OF PLATES

LIST OF FIGURES

LIST OF TABLES

LIST OF CONTRIBUTORS

R.K. Banerjee
National Atlas and Thematic Mapping Organisation
Calcutta

M.G. Chandrasekhar
National Natural Resources Management System,
Indian Space Research Organisation,
Bangalore

J.T. Coppock
Department of Geography,
University of Edinburgh,
Edinburgh

S. Jonna
National Remote Sensing Agency,
Hyderabad

J. Krishnamurthy
National Natural Resources Management System,
Indian Space Research Organisation,
Bangalore

M. Kudrat
Regional Remote Sensing Service Centre,
Indian Space Research Organisation,
Dehra Dun

A.K. Kundu
National Atlas & Thematic Mapping Organisation,
Calcutta

B. Manikiam
National Natural Resources Management System,
Indian Space Research Organisation,
Bangalore

Jack Massey
Space-Time Research,
Melbourne

D.K. Mittra
National Library,
Calcutta

P. Nag
National Atlas & Thematic Mapping Organisation
Culcutta

A.S. Padmavathy
National Natural Resources Management System,
Indian Space Research Organisation,
Bangalore

Haushila Prasad
Department of Geography,
Kenyatta University,
Nairobi

B.K. Roy
National Atlas & Thematic Mapping Organisation,
Calcutta

D. Saha
National Atlas & Thematic Mapping Organisation,
Calcutta

S.K. Saha
Indian Institute of Remote Sensing,
Dehra Dun

Smita Sengupta
Department of Geography,
University of Calcutta
Calcutta

J.S. Sharma
Regional Remote Sensing Service Centre
Indian Space Research Organisation,
Jodhpur

Narvada Sharma
Kendriya Vidyalaya,
Khetri Copper Complex,
Khetrinagar

S.K. Subramanian
National Remote Sensing Agency,
Hyderabad

R. Thiagarajan
Ministry of Science & Technology,
Government of India,
New Delhi

A.K. Tiwari
Regional Remote Sensing Service Centre
Indian Space Research Organisation,
Dehra Dun

A BRIEF PROFILE OF

DR. G.K. DUTT

Dr. G.K. Dutt has been a symbol of efficacy, efficiency and integrity. His contribution in the field of thematic cartography in India is significant. He always had a very clear perceptions of the goals to be achieved in all the assignments entrusted to him. He has been considered as a pole around which different activities related to thematic cartography have been initiated and flourished. It was due to his futuristic vision that the National Atlas & Thematic Mapping Organisation (NATMO) multiplied its activities in all directions.

Dr. Dutt (26-10-1929) was born in an educated and cultured family of geographers in the holy city of Varanasi. His elder brother, Shri P.K. Dutt, who died at a very early age, is still known for his contributions in geography. Dr. Dutt obtained B.Sc. degree from the Banaras Hindu University in 1950. Later he proceeded to England and obtained B.Sc. (Hons.) degree from London University. During his studies in Queen Mary College he came in close touch with the then celebrated geographers like late Prof. A.E. Smailes, Prof. S.W. Wooldridge, Prof. L.D. Stamp and Prof. J.F. Kirkaldy. Doctorate degree in geography was conferred on him by the Ranchi University.

On his return to India in 1958, he came in close touch with Prof. S.P. Chatterjee who was then trying to compile the English edition of the *National Atlas of India*. Dr. Dutt joined as a Senior Research Assistant and subsequently became the first Technical Officer and Joint Director and finally the Director of NATMO. Prof. Chatterjee could realise the potentialities of the young Dr. Dutt and keeping in view his training and background entrusted him the task of preparing the physiographic map series of *National Atlas of India* at 1:1 Million scale – a task which was then considered formidable. His contributions in other mapping projects were also unique. Due to his interest and position in the Organisation, the responsibility of designing, formulating and planning came naturally to him. He took a leading role in the foundation of the fourth, fifth, sixth and seventh plans of NATMO. The *Tourist Atlas of India*, the *National Atlas of India* in 8 volumes, the *Irrigation Atlas of India* and the like are the outcome of the initiatives taken by him at early stages. Large scale thematic mapping and urban studies were also introduced by him which now have become a major field of activity of the Organisation.

The thrust was not purely on mapping projects *per se*. The approach was for the all-round development of the Organisation. Popularising maps, organising seminars, computerisation in data storage, processing and analysis, remote sensing, training, scribing, digital mapping, participation in international exhibitions, joint projects with other departments, in-house printing are some of the areas where serious efforts were made. The seventh plan of the Organisation is the testimony of the consolidated endeavour made in the development of thematic cartography keeping pace with the modern times. The establishment of (a) Digital Mapping System, (b) Thematic Atlases Division, and (c) Research, Development and Training Division were mooted in this plan which will be considered as a major breakthrough in the Organisation's activities in the years to come. His vision of the future shape of the Organisation and its possible role in the nation-building was very clear to him. The monograph entitled *Scenario of the National Atlas & Thematic Mapping Organisation in 2001 A.D.* reflects his views.

In the international fields of cartography and geography Dr. Dutt's contributions have been quite significant. He has attended several international conferences and performed a leading role in organising the 21st International Geographical Congress in New Delhi in 1968. Later as the Director of NATMO he was instrumental in holding the International Seminar on Environmental Maps and Atlases and the International Conference on Population Mapping. He was the member of the IGU/ICA Joint Working Group on Environmental Maps and Atlases. At present he is a Commission Member of ICA Commission on Thematic Cartography from Satellite Imagery.

Dr. Dutt is closely associated with several organisations in the country. He is Vice-President of the Geographical Society of India and the Institute of Landscape, Ecology and Ekistics; General Secretary of the School of Fundamental Research, Calcutta; and President of the Confederation of the West Bengal Mountaineers and Trekkers. He has been member of the board of studies and examiner of several universities and expert in several committees and bodies appointed by the Government. He is a subject specialist in geography with the State Book Board, West Bengal.

There is a long list of publications which reflects his life long commitment to thematic cartography. Dr. Dutt is a softspoken and hard working person. He has the ability to challenge the issues which he does not consider appropriate.

Dr. Dutt will continue to be identified with the process of the modern developments in thematic cartography in India. The U.S. Bureau of Census communicates the following:

> "The Geography Division of the United States Bureau of the Census commends Dr. G.K. Dutt on his exceptional leadership of the National Atlas and Thematic Mapping Organisation of India, for a career of outstanding contributions to the discipline of thematic mapping. The cartographic products prepared during Dr. Dutt's tenure as Director ranks among the finest materials of this type in the world today. As active members of the cartographic profession, we are proud to recognise Dr. Dutt among its leaders, and offer our best wishes on this occasion of his retirement."

(Received by Robert W. Marx, Chief, Geography Division, Bureau of the Census).

1

THEMATIC CARTOGRAPHY : Commitments, Concerns and Solutions

PRITHVISH NAG & H. PRASAD

Books on thematic cartography are rare. National Atlas & Thematic Mapping Organisation (NATMO) produced an edited volume by Prof. S.P. Das Gupta entitled *Essays in Thematic Cartography* on the occasion of its Silver Jubilee (1981). It gives an idea about the issues in thematic cartography relevant during that time. The issues were related to (a) mapping techniques, (b) techniques of photo interpretation, (c) geomorphological mapping, (d) land use mapping, (e) mapping for development planning, (f) atlas cartography, and (g) place names in cartography. Regarding technological innovations, aerial photo and visual interpretation of the satellite imagery were considered the latest tools. Since then, almost a decade has passed. There has been a lot of technological innovations during this period. Digital remote sensing has come in a big way. Geographical information system (GIS) has become a catchwork in the scientific literature related to surveys, mapping and geo-sciences. Digital mapping with powerful softwares have been knocking at the doors even in developing countries like India. There has been a tremendous jump in technology which has influenced the thematic cartography as well. A thematic cartographer with training will now have to keep abreast with the fast changing technological scenario. The present volume looks into the pattern of development, the current issues and concerns in thematic cartography, the technological impact and possible development in thematic cartography, particularly in India.

History of Thematic Cartography

In his chapter on the development of this specialized field of cartography, Mittra has explained the roots of this subject. He had the advantage of his attachment with the National Library in Calcutta. Presumably the original source materials were available, based on which he has written his chapter. It is interesting to observe how the issues related to thematic cartography varies over time. The earlier maps were providing broad physical details of the country, while the recent maps are concerned with the resources, their assessment and management.

One of the off-shoot of this field of specialization is of course the *atlas cartography*. Kundu, Nag and Saha in their chapter studied the development of atlas cartography with particular reference to the *National Atlas of India*. The concept has been changing fast from the time of Prof. S.P. Chatterjee. The scope and coverage is likely to undergo drastically changed with the advent of superior technology (Dutt, 1983). *National Atlas of India* is being prepared on continuous basis only in India. Hence, this experience in a developing country has become a model.

Management of Thematic Information

Thematic cartography involves collection, compilation, processing and plotting of a variety of data ranging from statistics of remote sensing. Having a long experience of statistical data handling and preparation of thematic maps, Roy has highlighted the role of geographers in thematic cartography. He has traced the development of this field in several countries in comparison to India. The most important aspect of his study is the application of statistical thematic cartography in regionalization.

All sorts of information can now be procured and digitally overlayed with the help of the GIS technology. Powerful hardware and software can capture the information from a variety of sources and can process as well. The processed information can be used for report generation, tables, diagrams and maps in different forms, such as on paper, CCTs, CD-ROM, microfilm, etc. The management of information is now considered as a part of a larger exercise, such as land information system (LIS) or geographic information system (GIS). The former deals with information at a large scale, while the latter is more concerned with relatively smaller scale. Both the systems are complementary to each other and are used for different applications, such as land management, property records, urban planning, local administration and regional planning, resource management, and environmental protection (Dale & McLanghlin, 1988). In her paper on GIS, Smita Sengupta has dealt all these aspects at a greater length including the rôle of remote sensing data as one of the most vital source of current information.

Remote Sensing

With in last two decades, the scenario related to remote sensing has witnessed rapid changes. The data provided by each satellite and sensors have been better, particularly considering the resolution. The pixel size in the LANDSAT 1, 2 & 3 was 79 × 79 m which reduced to 10 × 10 m in SPOT panchromatic. However, LANDSAT Thematic Mapper (TM) is considered to be more versatile because it provides data in seven bands, including the thermal band. The India Remote Sensing Satellite IRS-IA has two sensors with the pixel size of 36.5 × 36.5 m in LISS II and 73 × 73 m in LISS I which are close to LANDSAT 5 MSS and TM (Table 1.1). The data from these censors are available from the National Remote Sensing Agency (NRSA) in Hyderabad.

The successful launching of IRS-IA was a boost to remote sensing application activities in India. Regional Remote Sensing Service Centres (RRSSC) were established in different parts of the country for providing facilities to users in processing the data. These centres were a part of a greater activity associated with the development of the National Natural Resources Management Systems (NNRMS). RRSSCs have powerful computer system for the manipulation of the remote sensing data, and are capable of identifying and mapping of different themes, such as water resources, forests, landuse, soils and the like. Manikiam *et al* have discussed the role such system in thematic mapping. It appears that for the mapping of natural resources based themes, NNRMS has not been fully exploited.

Another area of development related to remote sensing has been the instrumentation. Indian Space Research Organisation (ISRO) and its different units gave due attention to this vital aspect. Krishnamurthy *et al.* in their chapter enlisted several visual interpretation instruments, such as optical reflecting projector, diazo printer or optical pantograph. The licences of such instruments have been passed on to private companies as a part of a programme of technology transfer to private sector.

TABLE 1.1: **Features of Different Remote Sensing Satellites**

Features								
		LANDSAT (LS) 1,2,3		LANDSAT 4,5		SPOT	IRS-1A	
ORBIT DETAILS								
Altitude		918km.		705 km.		813 km.	904 km.	
Inclination		99°		99°		98°6'	99°	
Repetition Cycle		18 days		16 days		26 days	22 days	
Orbits		252		233		—	307	
Local time of passing equator		9.42 a.m.		11 a.m.		10.30 a.m.	10 a.m.	
IMAGING SYSYEMS								
Sensor	KBV	MSS	MSS	TM	HRV 1	HRV 2	LISS I	LISS II
Type of Scanning	Video cameras	Oscillating mirror Scanning			Push Broom		Scanning	
FOV (Scan angle)	—	11.56°	14.9°	14.9°	4,13°	4,13°	9.4°	4.70°
GRE/IFOV	40×40m	79×79m	79×79m	30×30m 120×120m	20×20m	10×10m	73×73m	36.5×36.5m

(Contd.)

TABLE 1.1 *(Contd.)*

Features		LANDSAT (LS) 1,2,3		LANDSAT 4,5		SPOT	IRS-1A	
Swath (km)	185 (251,2) 99 (253)	185	185	185	60	60	148	74×2
Frame Area	185×185 (251, 2)	185×185	185×185	185×170	60×60	60×60	148×148	74×74
Coverage	99×99 (253)							
No. of bands	Ls 1,2,3 Ls 3 : 1	4	4	7	3	1	4	4
Spectral Bands (Microns)	0.475-0.575 0.58-0.68 0.69-0.83 (Ls 1, 2)	0.5-0.6 0.6-0.7 0.7-0.8 0.8-1.1	0.5-0.6 0.6-0.7 0.7-0.8 0.8-1.1	0.45-0.52 0.52-0.60 0.63-0.69 0.76-0.90 1.55-1.75 2.08-2.35 10.4-12.5	0.5-0.59 0.61-0.68 0.79-0.89	0.5-0.73	0.45 0.52 0.62 0.77	0.52 0.59 0.68 0.86

Source: K.V. Ravindran, RRSSC, Dehra Dun.

Regarding digital image processing instruments, ISRO (SAC) development a standalone system called ISROVISION. It is a low cost system with a high resolution image display processor and comprehensive capabilities for image data handling, preprocessing, enhancement, information extraction and formatting functions. It is configured around a IBM-PC/AT-386 based host and a special purpose image display processor called VIBGYOR-2001. SACIMAGE software library is associated with this system. Other details are included in Table 1.2. This system is continuously being improved and within next 3 years an advanced version is likely to be brought out.

TABLE 1.2: System Specification of Isrovision

(I) HOST:

Hardware System Specification:

(i) CPU : 80 386 with 80387 co-processor
(ii) Memory : 8 Mbyte
(iii) Floppy Drive : 1.2 Mbyte (formatted)
(iv) Disk : 2 × 40 Mbyte
(v) Console : EGA

Option - 1 Cartridge Tape Drive (CTD).

Software specifications:

(i) Operating system :Xenis System V
(ii) Language support : Fortran compiler, C Compiler

(II) Application Package : SACIMAGE Library

(III) VIBGYOR - 2001

Input/Output Processor Specifications:

(i) CPU : Z - 80A
(ii) Memory : 64 k byte (RAM, ROM)
(iii) 1/0 controller 1/0 interface for Magnetic Tape Controller

Option - Streamer Tape Drive (STD)
Cipher Tape Drive
No. of Tracks = 9
Density = 1600/3200 cpi
Speed = 25 - 75 BPS

(Contd.)

(IV) Image display processor :

(i)	Image Memory Size : 1024 × 1024 × 8 bits	
(ii)	No. of Image Blocks : 16 of 512 × 512 × 8	
(iii)	Graphics plans : Any one out of four image planes	
(iv)	Lock-up tables : 3 × (256 × 8 bits)	
(v)	Display functions :	200m by 2 004 Roam over 512 × 512 cursor (Controlled by mouse)

Courtesy : A.R. Dasgupta, Space Applications Centre, Ahmedabad.

Unfortunately, despite all these efforts, ranging from the designing of indigenous satellites to the manufacturing of digital image processing instruments, the thematic mapping has not taken the advantage of the advanced technology easily available locally. Most of the thematic maps continue to be based on conventional sources of information, such as fieldwork, maps and statistics. On the other hand, the same satellite data can be processed by different methods in order to extract information on a variety of themes. Kudrat and Saha have highlighted the software options for such manipulation of the remote sensing data. The initial cost of data is expensive but one time investment for a good number of themes. Remote sensing data also forms an essential part of the GIS activity which is so vital for thematic mapping. With the advancements of digital technology, remote sensing has also come closer to digital mapping.

Digital Mapping

The development of digital mapping technology has been based on the efforts made in the complimentary areas of research and application. Some companies have developed this technology as an extension of digital remote sensing, while for some companies the experience is in the opposite direction. Further, the systems which were meant for graphics, shifted to CAD/CAM and finally to digital mapping. Similarly, the systems meant for engineering drawing were modified in order to serve the purpose of digital mapping.

Digital mapping is suppose to do the same work as done manually earlier. But, it has certain advantages as well. Such advantages are more significant in thematic cartography because the scope of experimentation and manipulation of data becomes enormous. In general digital mapping has following basic functions:

(a) Data capture in digital form,
(b) Data processing,
(c) Data presentation, and
(d) Transmission of graphical representation from digital to other forms.

Digital mapping functions have some degree of similarity with GIS activities. Both deal in one way or other with data acquisition, classification, structuring and restructuring, editing, generalization, enhancement and analysis. The basic issue in this method is selection of the mode of data acquisition and subsequent operations. The comparison of raster and vector modes has been shown in Table 1.3. However, programme for vector to raster conversion and *vice-versa* is now easily available. The raster data captured by scanner is faster in comparison to vector data acquisition by manual digitizer.

TABLE 1.3: Raster and Vectors

Feature	Raster	Vector
Data capture	Fast	Slow
Data Volumes	Large	Small
Graphics	Medium	Good
Data Structure	Simple	Complex
Geometrical accuracy	Low	High
Linear network analysis	Poor	Good
Area/polygon analysis	Good	Poor
Combination of data layers	Good	Poor
Generalization	Simple	Complex

Source : Dale & McLanghlin, 1988.

Mainframe based digital mapping solutions are being proposed by large multi-national companies. These systems are expensive and require specific environmental conditions and trained manpower which is not always easy to arrange in developing countries. However, PC based solutions are now available which suits the conditions that exists in such countries. The PC based software can almost do the same work as the mainframe based solutions. Further, there are solutions which are compatible with mainframe as well as PCs/micro-computers. For example, ARC/NFO solutions are available for both the systems. It can do almost all the GIS functions which Sengupta has mentioned in her chapter. Further, GIMMS, Edinburgh based software has been used on mainframes till recently. Now a PC version is also available which will require following hardware capabilities:

(a) IBM PC/AT or upwards (i.e. 80286 or 80386 processor)
(b) Floating point co-processor (i.e. 80287 or 80387 co-processor)
(c) At least 3 Mbytes of memory (extended memory for MS - DOS)
(d) MS-DOS 3.3 (or later) or OS/2 operating systems
(e) Hard disk with at least 5 Mbytes available space
(f) Parallel part with 25 pin D connector. (*International Journal on GIS*, 1989 Vol-3, No.2)

GIMMS software is good for GIS and digital thematic Mapping. It is designed primarily to present thematic data through a combination of map display, graphical and tabular output. GIMMS creates a variety of permanent files of information and these have a common interface to the operating system (Braclan & Weber, 1989). This is a user friendly package.

PC based solutions have already made an impact in cartography. Once data captured either by vector or vaster methods open new vista for applications contributing to the development of thematic cartography. One of the example is the introduction of personal computer (PC) and compact disc-read only memory (CD-ROM) technologies. The combination of both the technologies provide solutions for easy access of massive data sets, such as censuses. It is also possible to plot such data on the cartographic base captured by the vector method. With the spread of PCs all over, the solution for massive data handling, processing, and, plotting will be easy. In case a plotter is also attached to the PC/CD-ROM, coloured thematic maps can also be produced.

Thematic mapping has been considered as an of-shoot of geography. The technological impact has forced to revise geographical concepts, and models. Obviously, such an impact on thematic cartography is more explicit. Sharma and Sharma in their chapter have looked into these aspects.

Methods of Thematic Maps Reproduction

The conventional method of map reproduction is on paper which involves painstaking cartographic work and multicoloured printing. The printing technology is changing fast but a huge infrastructure is required. It is now possible to generate several shades of colours of tints which are extremely important for thematic mapping. But, printed thematic maps are becoming uneconomic due to the cost of paper, printed, maintenance of the infrastructure, transportation and postage. There has been attempts to reproduce the thematic maps on non-conventional mediums in order to circumvent the above problems. Advantages of the developments in optical and digital technologies, or combination of both have been taken in this regard. As a result thematic maps, though few, are now produced on microfiche, videodisk, CD-ROM disc CCTs. The advantages and limitation of these mediums have been cited in the chapter by Coppock. However, in India these methods are yet to be tried.

Emerging Issues in Thematic Cartography

Technology has become as essential component in the development of thematic cartography. As it will be apparent from the following chapters, the merging technologies will greatly influence this specialized field of cartography. Remote Sensing, computer, data capturing and printing technologies are closely associated with thematic cartography. Development in one sector is bound to influence this branch of cartography.

In the chapter by Thiagrajan, the future scenario of computers has been estimated. The storage circuits, nature of inputs and processing etc. will contribute to several options and alternatives in the choice of technologies in thematic cartography as well. The capacities of the host computer will have implications in remote sensing, GIS and digital mapping.

The technological support has already lead to the global thematic mapping activities on one hand (vide Coppock's chapter); and extension of national data generating system for thematic mapping on the other hand (vide Manikiam *et al*). Based on the possible technological options and thus likely merging patterns, stages for jumps in technology have been identified by Nag while building up the scenario for thematic cartography in India.

With improved technology, new challenges in thematic cartography are likely to be taken up. Maps involving voluminous data processing would be possible. Parallel processing of different data sets will reduce the time required for experimentation. New themes, considered near - impossible till recently, are likely to be accepted for mapping. Theme specific software will be developed and arranged in a batch mode. Thematic cartography will become a very specific field of specialization. Obviously, a thematic cartographer will have to undergo, several yeas of training in order to become an expert in this field. Thematic cartography is likely to become technology specific by the turn of century.

REFERENCES

Bakker, Nico J., Corne P.J. M. Van Elzakker and Ferjan Ormerling (1987), National atlases and development, *ITC Journal*, No. 1, pp. 83-92.

Bracken, I, and Chris Webster (1989), Towards a typology of geographical information systems, *International Journal on Geographical Information Systems*, Vol. 3, No. 2, pp. 137-52.

Das Gupta, S.P. (ed) (1982), *Essays in Thematic Cartography*, National Atlas & Thematic Mapping Organisation Monograph 2, Calcutta.

Dutt, G.K. (ed) (1983), *Atlas Cartography*, National Atlas & Thematic Mapping Organisation Monograph 3, Calcutta.

Dale, Peter F. & John D. Mclaughlin (1988), *Land Information Management*, Clarendon Press, Oxford.

Gutsell, B.J. (ed.) (1979), *The Purpose and Use of National and Regional Atlases*, Cartographica Monograph 23, University of Toronto Press.

Nag, P. (ed) (1984), *Census Mapping Survey*, IGU Commission on Population Geography, Concept Publishing Company, New Delhi.

Rhind, D.N., I.S. Evans and M. Visvalingam (1980), Making a national atlas of population by computer, *The Cartographic Journal*, Vol. 17, No. 1, pp. 3-11.

2

DEVELOPMENT OF THEMATIC CARTOGRAPHY IN INDIA

D.K. MITTRA

A map is a symbolic drawing of the visible as well as locational and distributional patterns of the whole or part of the earth drawn to scale. Whether it is made for general or any specific purpose, it communicates with the user of the map. In fact, it is a distinct form of graphics in the sense that it specialises in communicating facts and ideas about the earth. Hence, the success of a map is often determined by the effectiveness with which it is able to convey the message. From this point of view, a map, as a means of communication, communicates as the source material to the map user. In order to reach its destination, i.e. map user, a map has to pass through various channels. The cartographer, with his basic knowledge of the science of drawing maps, acts as a transmitting medium, recording various information with letters and symbols under particular arrangements. Sometimes even a finally drawn map, due to bad reduction and printing, often acts as a bad communicator. There are various other factors affecting communication by maps, e.g. projection, scale etc.

Early Maps

In the recent years, due to a great demand for maps by their users, the role of maps as a communicating medium has assumed importance. Early maps of India were based on verbal information supplied by travellers and sailors and for their use. They were more interested on broad details of the country and, such, the hills, rivers and main cities were incorporated without any stress on their size and shape. Ptolemy's map of India, Tabule Asiae X, as copied by Munster in 1540, displays the River Ganga rising from Imaus mountains

(Himalayas) and flowing into the Sinus-Gangeticus (Ganges Bay) along with the main mountain ranges. Giacomo Gastaldi's maps (1548 edition of Ptolemy) shows in detail the east and west coast towns of southern India, rather than the northern half.

17th Century Maps and After

A new trend of sea charts appeared in the 17th century as the frequency of voyages into and across the Indian Ocean continued to increase. The need for information was felt by the cart makers and they marked the direction of wind in particular seasons by arrows, indicating faster and easier sailing. Later, the charts included the eastern and western coasts and some of them even gave small views of the land as well, e.g. Ioan Jansson in 1658, Hendrick Doncker in 1665, Francois Velentyn in 1726 and Edmund Halley in 1728. The sailor could thus locate his position on the sea by comparing the coastline on the chart with the view from the ship.

Father Monserrate's map of India (c 1590) had general information about north and west coasts while Father Bouchet's map (1714) showed political divisions of South India. Both these maps varied greatly in terms of information content and area coverage. *Carte del 'Inde'* of d. Anville (1752), although showed rivers and mountains, kept much of the area as a blank, giving an indication that he was more interested on authentic information which would communicate a better picture of India. This is clearly evident from his two large scale maps, viz. Carnate and Coromandel (1753) — the south east coast of the country. This method also helped him to set down information in detail such as the note : "this is the part of India where the settlements that support the trade of the Europeans. . ." The aim and importance of maps soon changed after the battle of Wandiwash (1760) when they were consulted on a more scientific basis by the army. While traversing the country, the army started taking various readings, later transferring them a large maps.

Such a system of mapping gradually changed and, when Rennell came, a more systematic scheme was drawn up. The maps which were, till then, mainly meant for military campaigns, lacked details of the whole area. With Rennell started a systematic survey to facilitate collection of revenue and for transport and for communication. His *Map of Hindoostan* (1782) still remains the basis of Indian cartography.

Early 19th Century

By the close of the 19th century, knowledge of the land was increasing so rapidly that it had become impossible to complete a map of India by Rennell's method. By 1802, survey began to assume a scientific character with the measurement of Lambert's Madras base line. *The Atlas of India* was planned in 1824 (1" to 4 miles) to cover the whole of India and sheets were published over a period up to 1867 and later. In the meantime, from 1847 till 1883, the topographical, revenue and cadastral surveys were taken up on a reproduction scale of larger than 1" to 4 miles. However, there were no systematic plans. It was 1883 onwards that the topographical maps (1" inch and larger) were standardised. The Survey of India, up to date, is primarily responsible for topographical surveys and for the preparation of up-to-date maps covering the whole country on standard scales. In recent years Survey of India has also published pre-investment and river valley project maps.

Apart from the topographical maps, thematic maps are also being prepared. While the first category caters to the need to the general map users, the latter are for the specialists who depend upon various information. Although maps of the second category have a long history, it had been first evolved and pragmatised in India by Prof. S.P. Chatterjee in the form of the *National Atlas of India* in 1957. This had led other institutions to prepare thematic maps and atlases in the country.

Census Atlas of India (1961) is a work on the national level, based on census and extra-census data covering 194 maps. The majority of these maps are in black and white, while a few are in colours. Yet another work at this level on natural vegetation and agriculture has been taken up in 1962 by the French Institute, Pondicherry, in collaboration with Indian Council of Agricultural Research, New Delhi. The co-operation involves the Government of France and the Government of India.

At the state level, some work has been done with different accomplishments. Soon after attaining independence, two main atlases were produced in India in 1949. The first being *Bengal of Maps* by Prof. S.P. Chatterjee and other being *Economic Atlas of Andhra Pradesh* by Ramanadhan and V.L.S. Prakasa Rao. Both these atlases were prepared at a time when new state boundaries had either been or were being demarcated. This was quickly followed by a succession of specialised maps in atlas form, e.g. *Atlas of Livestock and Livestock Products*, 1950, and *India in Maps*, 1950. Both these were produced by the Government of India. Similar to Chatterjee's *Bengal in Map, Bihar in Map* by P. Dayal 1953 was another attempt in this direction. There were some more state atlases dealing with agriculture, power etc. mainly by graphs and charts.

1960 and Afterwards

Strictly speaking, from 1960 onwards thematic cartography has been a very high order. *Planning Regions in the Mysore State : The Needs for Re-adjustment of District Boundaries* by V.L.S. Prakasa Rao and L.S. Bhat (1960) and *Mysore State : An Atlas of Resources* by A.T.A. Learmonth and L.S. Bhat are two of the finest examples of thematic cartography. *Economic Atlas of Madras State* by National Council of Economic Research came out in 1962.

Nearly all states produced their own state census atlases, aggregating to 169 maps, mostly in black and white. Their content and presentation were of the same types, similar to the *Census Atlas of India*. This has been followed by an *Administrative Atlas* at the state level, but only Orissa could produce it effectively.

Census atlases of the states and of India cover mostly physiographic conditions, demographic structures and trends, agriculture, industries, employment conditions, castes, tribes, religions, education, housing and health conditions. *National Atlas (Rashtriya Atlas in Hindi)* gives a general picture of the broader regions of India with the physical background and socio-economic structure and throws light on the lines of future development of the country. The projection used for the maps of the entire country or major parts thereof was Lambert's Conformal Conical which maintains a good angular relationship while distortion is relatively small within latitudes 8° N and 36° and longitudes 68°E and 96°E. Most of the maps drawn on the scale 1:15m or larger have colour tints, colour screens or point symbols and proportional circles. These representations, along with the suitable projection, has made possible for the maps to communicate all that have been incorporated therein for depicting India's condition. This mode of communication has not been successful in the map of rural population with 15 density categories. While the *Census Atlas* has been drawn on the same scale as the *National Atlas of India*, cartographic representation has been simple but the importance lies in the fact that such an atlas was produced for the first time which fact itself is of great value. The vegetation map published by French Institute, Pondicherry on a scale of 1:1m shows very clearly the natural vegetation on different sheets (covering nearly 16,000 km^2 on each sheet) showing three major groups and 32 sub-types of vegetation. These maps are very useful as they give an idea of the potentialities of the region and of the chances of success in the introduction of new crop or afforestation of species.

NATMO and the Recent Scene

National Atlas and Thematic Mapping Organisation, Government of India has lately published *The Irrigation Atlas of India, The Tourist Atlas* and *The Atlas of Forest Resources,* proving that thematic cartography has taken a deep root in India with continuous experimentation and that of a better representation of data is being incorporated in new atlases. The *Irrigation Atlas* contains maps on food crops, Industrial crops, rural-urban population, to show irrigation and its allied aspects in a most detailed and systematic arrangement of facts. It is the first efforts of the kind in this country. It may, however, be interesting to note that *Atlas on Water Resources Development Atlas* is under preparation with 1:1m and 1:2m scale maps having information of surface water, runoff, utilisation and demand of water. As this atlas will be utility oriented, it will be very helpful in planning water management.

The *Tourist Atlas of India* includes the new edifices of modern India, especially the gigantic river valley projects and industrial complexes, as well as the places of art and culture and archaeological remains. For the first time tourist and allied information is available in detail.

The *Atlas of Forest Resources of India* comes under resource atlas series. This atlas, being the first of its kind, covers interesting information on forests and various economic products which are derived from them. The maps on soils and the type of rocks on which forests develop can be useful for practising foresters and silviculturists and the atlas as a whole will be helpful in the scientific management of the eco-system.

The same organisation has recently taken up mapping of land use and landform of selected problem districts of the country as it was thought to be appropriate to take up topics related to the nation's land resources and its characteristics as the exist in the country. The maps will be on scales of 1:250,000 and 1:50,000 covering the districts. Information will be acquired by air-photo supplemented by detailed field verification. Information on land use and landform are complementary to each other and are of fundamental importance for any type of land-planning or development-planning. There is no doubt that these maps would prove to be very effective for formulating future planning programmes.

The important urban areas of the problem districts would also be studied with a view to assess the relationship of such centres with their surroundings and to determine their growth potential. Thus these three types of basic information in the form of maps and monographs for each of the districts would be real authentic sources of information for a variety of purposes. However, the *Atlas of the City of Calcutta and its Environs* was prepared for a different purpose. It was NATMO's contribution on the occasion of the tercentenary of this great city (Kundu & Nag, 1990).

Presage

Thematic cartography has been a later development in the history of cartography. The case of India is not different. Government institutions, universities, firms and individuals are now involved in preparing atlases of all sorts. Preparation of loose sheets of thematic maps are yet to become popular. Considering a variety of applications the demand of such maps is likely to increase in the near future. Both the products will have their users. Thematic atlases will be more used for educational and reference purposes, while the loose sheets will be appreciated by the planners and resource managers. However, with the introduction of modern technology, there will be additional ways of producing thematic maps, for example in digital form. Thematic cartography in India awaits a quantum jump. By the end of the century the scenario is likely to undergo drastic changes.

REFERENCES

Bhat, L.S. and V.L.S. Prakasa Rao, (1960). *Planning Regions in the Mysore State: Th Needs for Re-adjustment of District Boundaries*, Indian Statistical Institute Calcutta.

Chatterjee, S.P., (1949). *Bengal in Maps*, Orient Longman, Calcutta.

Dayal, P., (1953), *Bihar in Maps*, Kusum Prakashan, Patna.

Gole, Susan, (1980). *A Series Early Printed Maps of India*, Jaya Prints, New Delhi.

Kundu, A.K. & P. Nag, (1990). *Atlas of the City of Calcutta and its Environs*, NATMO, Calcutta.

Learmonth, A.T.A. and L.S. Bhat, (1960). *Mysore State: An Atlas of Resources*, Indian Statistical Institute, Asia Publishing House, Bombay.

Lister, Raymond, (1970). *Antique Maps and their Cartographers.*

NATMO, (1957). *Rashtriya Atlas* (in Hindi), Calcutta.

NATMO, (1974). *Tourist Atlas of India*, (2nd Ed.), Calcutta.

NATMO, (1976). *Atlas of the Forest Resources of India*, Calcutta.

NATMO, (1980). *Atlas of the Agricultural Resource of India*, Calcutta.

NATMO, (1982). *National Atlas of India*, National Atlas and Thematic Mapping Organisation, Calcutta.

NATMO, (1989). *Irrigation Atlas of India*, (2nd. Ed.), Calcutta.

NATMO, (in press), *Water Resources Development Atlas of India*, Calcutta.

Phillimore, Col. R.H., (1945). *Historical Records of Survey of India*, Vol. 1, Geodetic Branch, Survey of India, Dehra Dun.

Ramanadhan, V.V. and V.L.S. Prakasa Rao, (1949). *Economic Atlas of Andhra Pradesh*, Vishakhapatnam.

Rao, V.L.S. Prakasa and L.S. Bhat, (1960). *Planning Regions in the Mysore State : The Needs for Re-adjustment of District Boundaries*, Indian Statistical Institute, Calcutta.

Raza Moonis and Aijazuddin Ahmad, (1990). *An Atlas of Tribal India*, Concept Publishing Company, New Delhi..

Roy, B.K. (Ed.), (1971). *Administrative Atlas*, Registrar General of India, New Delhi.

Roy, B.K. (Ed.), (1981). *Census Atlas of India*, Registrar General of India, New Delhi.

3

CHANGING CONCEPT OF NATIONAL ATLAS

P. NAG, A.K. KUNDU & D. SAHA

This chapter has been written with an aim to find out the possible changes in the thematic coverage of a national atlas, particularly with reference to India. The issues in thematic mapping have been continuously changing thereby influencing the contents in the national atlas which is a collection of thematic maps in an orderly fashion. The coverage of such a volume can no longer be static over time. This is not only due to the shifting focus of the national scenario, but also due to the versatility of the data now being made available with matching technology for processing, experimentation and printing. The topics at certain scale which were considered impossible for mapping, have now come within reach.

The earliest known reference of an atlas prepared in India is the *Bengal Atlas* by Sir James Rennell in 1781 followed by *Atlas of India* showing Provinces in the Bengal Presidency and Western Provinces which was published by the then British Government of India in Calcutta. Following this atleast one more atlas is known to be published by the same Government entitled *Statistical Atlas of India* prepared for the colonial and Indian exhibition, 1886 (2nd ed. in 1895). This publication was also from Calcutta. Before independence some attempts were made by some private publishers and the Survey of India to produce atlases. The most significant contribution in this direction was from Prof. S.P. Chatterjee. His *Bengal in Maps* was again published from Calcutta in 1949. Thus atlas cartography has got deep roots in Calcutta, where NATMO has been rightly located.

The circumstances leading towards the preparation of the *National Atlas of India* had something to do with the emergence of Prof. Chatterjee on the international scene. In 1956 he became a member of the International Geographical Union (IGU) Commission on National Atlases. Same year, at the instance of Pandit Jawahar Lal Nehru, the first Prime Minister of India, National Atlas Organisation was founded and in the following year the first *National Atlas of India* was produced in Hindi.

Since then several atlases on different themes have been produced. This commission continued up to 1976 which was succeeded by an IGU/ICA joint working group on Environmental Maps and Atlases headed by D.P. Bickmore who visited NATMO twice in last few years. Dr. G.K. Dutt, ex-Director NATMO, became a member of this commission, and through him contacts were maintained with this international scientific group. But the focus of this group diversified in order to include the preparation of the digital base for environmental mapping. Recently International Cartographic Association (ICA) has appointed a commission on National Atlases with the following aims (ICA, 1988):

- National reports: the current situation and production techniques.
- Technical systems and software: the current market and future trends.
- Electronic/interactive atlases: technical, educational and distributional aspects.

Changing Themes in National Atlas

Since the time of the first publication of National Atlas, the concept of national atlas is changing within the national framework or from country to country. Dr. W. Stamps of the Technical University Dresden, G.D.R. has identified the changing themes during the periods as follows:

Up to 1850

Atlases published up to the middle of 19th century did not usually contain thematic maps. Pure thematic atlases (referred as specialistic atlases) were extremely rare. Therefore, most of the atlases prepared during this period were of reference atlas type. These atlases are of historical importance.

1851-1980

During this period thematic atlases of 3 basic types were produced; the complex, semi complex, and the mono thematic atlases. This period is still determined by predominating traditional cartographic techniques which strongly influenced cartographic representation.

1919-1945

Within this period the number of regional atlases increased and at the same time the first planning atlas along with complex regional atlases appeared while the number of regional school atlases became frequent.

1946-1960

After Second World War the production of atlases revived which were mostly many thematic special atlases and semi complex atlases and rapidly more complex atlases were produced.

1965 onwards

During this period the publication of regional atlases, beginning of national atlases and production of complex regional atlases are the prominent features. Such activities were greatly influenced by the IGU Commission on National Atlases headed by Prof. Salichtchev.

There are very few countries who have prepared national atlas several times. The example of the Netherland is in front of us. The origin of the Dutch atlas goes back to 1932, but there was a set back due to economic depression and the war (Baker *et al.*, 1987). This atlas appeared between 1963 and 1977 containing 13 issues or themes or 98 loose leaf map sheets. Further, the supplements of 16 updated or loose map sheets were published during 1978-81. Themes in the second edition of the atlas changed in order to include more problem-oriented issues. For these, dynamic themes such as climate or soil, consultation of the first edition was required. The new atlas was quite different from the earlier first edition. Instead of loose sheets there will be 20 separate volumes. Each volume is in fact a small thematic atlas and resembles any standard geography textbook. This publication, started in 1984 is likely to be completed by 1988. The themes in this atlas are as follows:

(a) Population and settlement — (i) Population, (ii) History & Human Occupation (iii) Towns and Cities, (iv) Villages.
(b) Human Welfare — (v) Housing, (vi) Service, (vii) Recreation.
(c) Human housing — (viii) Labour, (ix) Industries, (x) Agriculture, (xi) Gateway to Europe, (xii) Infrastructure.
(d) Natural Resources and Environment — (xiii) Geology, (xiv) Soils, (xv) Water, (xvi) Landuse.
(e) Physical Planning — (xvii) Physical Planning, (xviii) Urban Planning, (xix) Rural Plan.

Out of the above 20 volumes, 12 volumes were already published by 1987. Further the above scheme provide us the scope and coverage of the latest Dutch national atlas.

Since the publication of the first national atlas of Finland in 1899, followed by Canada in 1906, a lot of changes have taken place concerning contents. In this 90 years' period, several editions of national atlas of these countries and other countries have been published. Though, all the themes of the editions of national atlases are not known, the information available indicates that focus, coverage and contents have been changing. The shifting focus is perhaps due to the compulsion and issues which gain prominance in respective nations at particular moment of time.

The movement for preparing national atlases has spread all over. Only 12 countries prepared their national atlases by 1960, which increased to 54 in 1978. For example, in Africa only Egypt and Tanzania produced their national atlases before 1960, and in the next 12 years Sierra Leone, Cote d'voire, Ghana, Nigeria, Uganda, Kenya, Zambia and Madagascar published their national atlases. It may be mentioned here that most of these countries got independence during this period (1960-78).

The Indian Case

The example of India is some what different from the experiences of other countries who are even having longer tradition of preparing national atlases. The most significant aspect is the organisational structure. Most of the countries, similar to India, took up projects on national atlas. Since each theme in such atlases require good scientific backing, in most of the countries expert committees were attached to a thematic map or a set of such maps. Obviously, when the map is printed, there is no role for such committees which may be having experts on data collections, designing, remote sensing, digital mapping, geography and cartography. In India, the situation is different. National Atlas & Thematic Mapping Organisation is a permanent institution established on government funds. There are no expert committees associated with particular theme. Hence, for each theme the expertise has to be developed within the organisation. If a new theme has to be taken up, the relevant scientific backing, in terms of expertise and infrastructure, has to be developed from within. Obviously, such preparations require advance planning. If the trend in developed countries or emerging issues in India is to kept in mind, more and more problem oriented thematic maps should be getting prominance in the forthcoming editions of the national atlases.

In the first edition of the *National Atlas of India*, there were 26 plates out of which about 11 themes are static in nature. This number can increase if we take the scale into consideration. The thematic contents in the maps tend to be less and less dynamic as the scale is reduced because the degree of generalisation is to be increased. Hence, the themes may appear to be dynamic, but due to scale the contents become static.

The contents in the second and enlarged edition changed grossly. The number of plates increased many folds, i.e. from 26 to 300. The scale of the common thematic maps changed as well. Cartographic base at 1 : 1m, 1 : 2m and 1 : 6m were introduced. For the first two scales, the country was covered by several plates — 15 for 1 : 1m and 5 for 1 : 2m series. The administrative plate which was at 1 : 5m in the first edition was changed to 1 : 1m in the following edition.

The second edition is divided into eight volumes in order to cover themes considered relevant for the country (Table 3.1). From the table it is apparent that the themes which were covered in 1 or 2 plates in the first edition are now covered almost by one of the 8 volumes of the second atlas. Further, out of the above volumes, the themes falling in the first 3 volumes are more-so-over static in nature. But in other volumes, the theme and scale makes individual plate static or dynamic. The classical example is of land use plates (157 – 171) at 1 : 1m scale and India – Land Use Plate (172) at 1 : 6m scale. The latter map is static in nature due to the scale.

TABLE 3.1: **National Atlas of India — Plates and Contents.**

Volume	No. of Plates	Important Themes
1. General Political	24	Adjacent countries, Administrative Parliamentary Constituencies.
2. Physical & Geomorphological	37	Physical Slope, Rocks & Minerals, soils Texture & Reaction, Earthquake.
3. Climatic & Biographical	48	Monsoons, Drainage, Water Resource, Wildlife & Wetlands.
4. Population & Transport	47	Population, Density, Transport & Tourism, Railways, Roads, Post & Telecommunication.
5. Landuse & Industrial	26	Landuse, Manufacturing, Chemical, Sugar Industries, Power Resources
6. Agricultural & Economic	42	Crop Regions, Irrigation, Rice, Wheat, Millets etc. Soil Regions, Limestone, Fisheries, Land Ownership.
7. Soil & Regional	45	Diseases, Literacy, Land Holding, Working Force, State Maps.
8. Historical & Cultural	31	Historical, Cultural Landscape, Village Types, Urban Landscape.

Source: *National Atlas of India*, 1982.

Influence of Thematic Atlases

Alongwith the preparation of the *National Atlas of India*, other atlases with a focal theme were also produced by NATMO. These thematic atlases were taken up either on special request by some government agencies or by our own assessment about its potentialities. In most cases, a major part of the data was supplied by the requesting agency. The example of such atlases are *Irrigation Atlas of India* (1972), *Tourist Atlas of India* (1974), *Atlas of Forest Resources of India* (1976), *Atlas of Agricultural Resources of India* (1981), and the *Water Resource Development Atlas of India* which is almost ready. Some of the themes in such atlases are either common or static in nature. But, a majority of plates have specific themes relevant to the thematic atlases. These plates initially prepared for these atlases found their way to the *National Atlas of India*. Conversely, the plates already planned for the *National Atlas of India* were modified in order to meet the requirements of the thematic atlases. Hence, a complimentary relationship has already developed between the *National Atlas of India* and the thematic atlases. Revision of common plates can benefit both the atlases.

Influence of other Cartographic Activities

Apart from preparing the *National Atlas of India*, in English, Hindi and regional languages, the other mandates for NATMO are as follows:

- Preparation of thematic maps based on research studies on environmental aspects and their impact on social and economic development.
- Preparation and compilation of land use and landform maps.
- Urban studies.
- Geographical researches.

The above activities are complimentary to the main objective of the Organisation. Any progress in one area of activity is likely to influence the thematic contents of the 300 plate volume (Fig. 3.1).

The conferences and seminars which are held almost regularly every year focus on the themes of great national importance of the particular moment of time. Such gatherings provide us opportunities to have an insight into our own activities and to have a dialogue with the greater scientific community particularly for taking up challenging themes for mapping. The monographs published on such occasions or otherwise are a sort of a permanent record for consultation. Similarly, for the land use, landform mapping and urban studies, recent information is collected which help to update the information base for the national atlas as well.

Influence of Distant Cartography

With all sort of innovations, thematic cartography in phases is increasingly becoming programmable. After initial contacts with conventional techniques in mapping, thematic maps can be prepared even by maintaining a distance. Digital image processing and mapping systems based on well programmed geographical information system has opened up new vistas in thematic mapping. Such advancements have permitted to take up new themes for mapping which were near impossible earlier. The temporal phenomena, change detections, environmental mapping and the like are now possible with the remote sensing data. The provision of obtaining data at regular interval is likely to bring a revolution in atlas cartography. Further, it is also becoming possible to manipulate these data sets to meet the spatial (projection) and thematic requirements for maps.

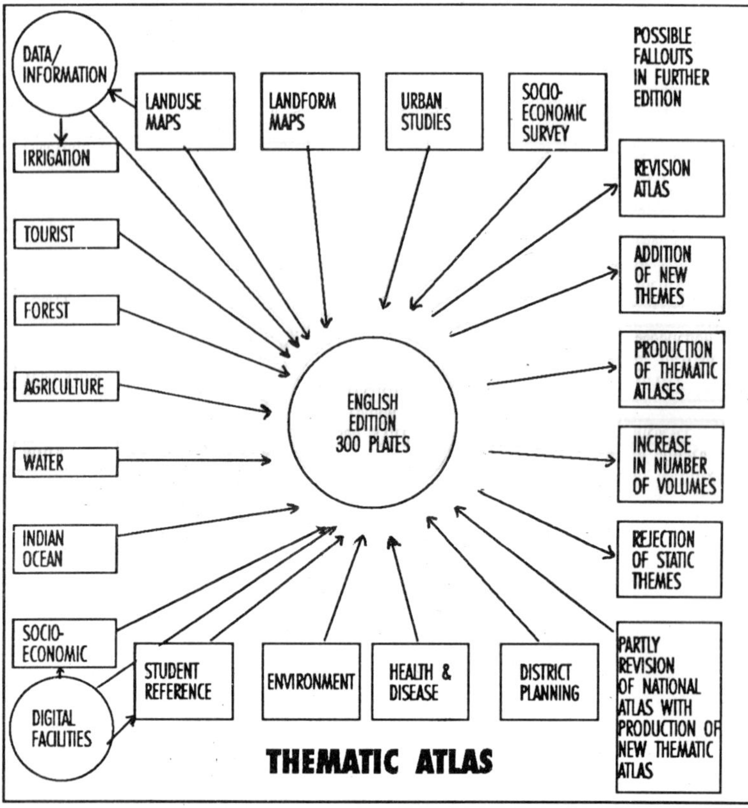

FIG. 3.1: Factors Influence in Changing Contents

The digital mapping system will allow to carry out a series of experiments for mapping new themes. The data sets from the diverse sources can be fused and a series of new maps can be prepared.

Options and Provisions

With the completion of the whole of the *National Atlas of India*, a revision cycle has already started either considering the demands of the users or due to preparation of the other atlases having common themes. Consequently, the maps are revised or reprinted. The time has come when a major policy decision has to be taken in this regard. What are the options available to us? How does our present and proposed activities fit into the various options open to us? What are the ways to make the whole institution financially viable? Some of these questions have been discussed in our occasional monograph entitled *National Atlas and Thematic Mapping Organisation in 2001 A.D.* (Dutt & Nag, 1987). Considering the themes related issues in mind, following options are available to us:

(a) *No Change in the content*

In the first option, the contents of the *National Atlas of India* can be kept same, perhaps with same number of plates (Table 3.1). With the new data set, the maps can be revised. The scales of the maps remain the same. A revision cycle could easily be worked out in this option keeping in view of the availability of the data.

(b) *Continuous addition of maps*

In this option, the number of plates go on increasing with addition of new themes in the national atlas. As and when, the themes which are considered significant for mapping can be taken up. Under such circumstances this atlas which is already bulky becomes of unlimited size, perhaps, the whole national atlas will never be available at one moment of time.

(c) *Fixed number of plates by substituting themes*

The number of plates can be fixed and themes which are static or have lost significance can be excluded. This will permit to keep on accepting and rejecting themes for the *National Atlas of India*. In this arrangement, the system of numbering of plates have to be worked out. The current themes which were undertaken as a part of a thematic atlas such as water resource, can be included. This system makes the national atlas more dynamic.

(d) *Division of the National Atlas into Static and Dynamic Volumes*

As already observed, the themes in first 3 out of 8 volumes are more static in nature. These volumes can be kept as such. For rest of the volumes a scheme of changing themes can be introduced. The number of plates in this option can be kept constant.

(e) *Continuous addition of volumes*

In this option any new thematic atlas, such as *Atlas of Agricultural Resources* automatically becomes a volume of the *National Atlas of India.* The number of volumes keeps on increasing. Preparation of this atlas becomes an unending process.

Prevision

National atlas can no longer remain a one time project work. Several countries within their set up are producing more than one edition. In every successive edition new themes are atleast added. With every new data set, the information is becoming more and more refined and detailed. This will have an impact on the selection of themes. Remote Sensing has been an additional source of input. Digital processing and manipulation of such data, it is becoming increasingly possible to take up new themes for national atlas mapping. In addition, the typical themes are likely to come up from time to time, such as resource mapping. National atlas, which has an important role in planning, cannot be aloof from the national priorities and policies. Hence, a national atlas can not be static in terms of the thematic contents.

What should be the process of shifting focus of the national atlas? How many themes or plates or maps should be changed in every new edition? What are the themes which are likely to be included keeping in view of the scenario perceived for the country in the 21st century? These questions are to be worked out in details if the national atlas is to be prepared, revised and updated on a continuous basis. On the other hand, there will be several options in chosing the technology to meet the challenges of thematic cartography. Such facilities will help in selecting a variety of themes. It appears that data, technology and funds will in near future permit to prepare a large number off thematic maps for the national atlas in any country, including India. Atlas cartography in India has perhaps matured enough to absorb technological shocks, thereby making a quantum jump in this field.

REFERENCES

Baker, Nico J., C.P.J.M. Van Elazakkerr and Ferjan Ormeling (1987), National Atlases and Development, *ITC Journal,* No.1, pp. 83-92.

Dutt, G.K. & P. Nag (1985), *Life and Work of Prof. S.P. Chatterjee,* NATMO occasional Monograph No. 1, Calcutta.

Dutt, G.K. & P. Nag (1987), *Scenario of National Atlas & Thematic Mapping Organisation in 2001 AD,* NATMO Occasional Monograph 3, Calcutta.

Elzakker, C.P.J.M. Van (1982), A Systematic Approach to National Atlas Mapping, *ITC Journal,* No.2, pp. 145-50.

Eutsell, B.J. (ed) (1979), The purpose and use of National and Regional Atlases, *Cartographica,* Monograph No. 23, Univ. of Toronto Prcss.

International Geographical Union (1988), Report on Thematic maps of the world, unpublished.

ICA (1988), *International Cartographic Association Newsletter,* No. 12 October, 1988.

Nag, P. *et al* (1982), Importance of inset maps and diagrams in Thematic Mapping in S.P. Dasgupta (ed), *Essays in Thematic Cartography,* NATMO. Monograph No.2, Calcutta, pp. 20-3.

Nag, P. (ed) (1984), *Census Mapping Survey,* IGU Commission on Population Geography, Concept Publishing Company, New Delhi.

Nag, P. & A. Dutt (1986), Issues in Environmental Mapping in H.H. Singh, P. Nag *et al* (eds), *Geography and Environment: Issues and Challenges,* Concept Publishing Company, New Delhi.

Nag, P. (1987), A proposed Geographical information system for India, *International Journal on Geographical Information System,* Vol.1, No.2, pp. 181-7.

Nag, P. (1992), Role of NATMO in Promoting Indian Geography, *National Geographical Journal of India,* 27 IGC special issue (submitted for publication).

NATMO (1982), *National Atlas of India,* Deptt. of Science & Technology, Calcutta (8 Vols).

Salictchev, K.A. (1972), National Atlases, *Cartographica,* Monograph 4, Toronto, Canada.

Smidt, M de. (1988), The human scope of Cartography — towards a new National Atlas of the Netherlands, *Cartographic Journal,* Vol.23, pp. 70-2.

Stamps, W. (1984), *National and Regional Atlases* — A Bibliographical Survey, International Cartographic Association Publication.

Yonge, E.L. (1957), National Atlases: A summary, *Geographical Review,* Vol.47, No.4, pp.570-78.

Yonge, E.L. (1962), World and Thematic Atlases: A summary survey, *Geo raphical Review,* Vol.52, No.4, pp. 583-96.

Yonge, E.L. (1962), Regional Atlases: A summary, *Geographical Review,* Vol. 53, No. 3, 47-32.

4

STATISTICAL THEMATIC MAPPING

B.K. ROY

The whole development of geography has been associated with cartography and most maps are the accumulation of experiences. However, maps become outdated as soon as they are printed off but, no doubt, these serve as one of the most powerful tools of technology for investigations and record of the present. Further, it is noted with concern that map making and thereby its products usually do not attract the users due to limitations of appreciation by them because sometimes it is strenuous mainly for those maps which are on large scales, for their arduousness of handling and lack of documentation on map patterns. Of course, "maps are the tools of Geographers" may not be usually the hypothesis but it should be manifestation of present economy, society and habitats (lands) which should be viewed by all concerned without any prejudice as it opens grounds for assessment and further investigations.

Some Developments

Thematic mapping has no rigid definition. It is an art to project through maps of various scales in such an integrated manner that the status of resources can be seen correlatively in its whole. Broadly thematic mapping can be divided into three components viz., (a) dealing with schematic mapping of resources by compilation where the correlation of statistical data may be difficult to match with the map. One of the best examples, is of the land use maps mostly of the non-statistical category, (b) a combination of statistical materials with the schematic trends to be exemplified by either a climatic map or a soil classification map, statistically interpreted in relation to climatic data and soil ingredients respectively and (c) thematic maps having direct relevance to transformation of statistics by statistical methods to probe into the dynamics and trends. In the later case, maps relating to a large number of themes such as population, economy, wealth, social dissimilarity assessments etc. are the concern. On the whole, thematic mapping tasks are difficult, complex and time consuming in processing and challenging as well.

In case of thematic maps generated on the basis of statistics or statistical methods which so far composed on small scales have evinced on the contrast some favourable positions by the users. However, the contents and themes on such maps have been mostly simple and direct in the form of diagrammatic representations. Gradually, due to advancement of statistical science, econometric formulations and stochastic processes have made quantification in such a complex situation that generating diagrams in traditional forms and recently by computer and digital input have made the task of cartographers more challenging. Some of these issues form the content of this paper to further develop the philosophy of the art of map-making.

Development in cartography in the western countries occurred due to firms producing maps with constant research inputs. To exemplify, it is quite known that George John Bartholomew (son of John Barthdomew) of the U.K., at the age of seventeen experimented hill shading by contours and layer colouring (1880) gave birth to this art to further develop the topographical and geographical cartography. On the other hand, statistical input to thematic cartography by Humboldt, based his map of isothermal lines in 1844. It is quite well known also that non-Geographers have contributed in most basic ways to enrich geography and cartography, e.g., Gardiner G. Hubbard, a Boston lawyer spirited and established the National Geographic Magazine (1888). This followed his son-in-law, Alexander Graham Bell, inventor of the telephone who succeeded Hubbard in 1898 brought innovations to the National Geographic magazine in producing a lot of materials for cartographic knowledge (Span, 1989). In Britain the Geologist Sir Dudley Stamp (late) later turned to geography created a landmark in thematic cartography, specially in land use mapping. In India, similarly, a geologist Padmabhushan S.P. Chatterjee (late 1988) turned geographer and cartographer and introduced *Bengal in Maps* (1949) and subsequently established the NATMO and produced the first *National Atlas of India* (1956) combining statistical thematic mapping in a major way, are some of the potential examples of creators' to cartographic knowledge most useful in broadening the horizons of millions.

In addition, several maps of the nineteenth century, specially from the censuses of Britain, Ireland and others attempted statistics oriented thematic mapping to probe into the distributions of population in several ways, giving more emphasis to colour cartography. At that time, the negative areas could not be viewed while plotting the distributions and densities; still it is missed in general on maps of small scales. The processes of experimentation and research on methodologies continued and various systems of thematic mapping took up a shape which exist today. However, the appreciation in using the maps are continuing at a modest pace. Cartographers may need not worry about this but may continue seriously to project by latest methodologies with innovations and this mission has to be enlarged continually on current problems.

Luckily, the developing countries in Asia and Africa are trying to project their problems such as on population, planning, resource utilisation and several alike in many dimensions through thematic maps. In India, fortunately, development of cartography both thematic and topographic are sponsored by the government as a well organised and specialised venture to help and study the present socio-economic levels. It is well known that the Survey of India is responsible for topographic survey related cartography, while the NATMO (National Atlas & Thematic Mapping Organisation) concerns with general thematic mapping programmes. The Census of India's Map Division works on the charter with special thematic mapping for the census and statistics related activities and produce one of the largest number of maps for country's decadal census programmes. In addition, the newly created NRSA (National Remote Sensing Agency) and its associated organisations are yet to start in a major way satellite cartography with special reference to resource inventories by transforming imagery digitised data. It is a healthy sign that firms and publishers on cartography are coming up in the country and it is hoped that all the combined efforts of the government and private enterprises and endeavours may lead to produce cartographic materials profusely to dessiminate knowledge of our resources, people and the country and an awareness may reach soon demanding lot of map materials by the public.

Present Issue

In the above backgrounds an attempt is made in this paper to project certain developments of thematic mapping in relation to statistics in the country with cross references of literature of other countries. As the nature of statistics is changing and *magnus* in output, the challenge to cartographers to cope up with the ever increasing statistics to transform thematic mapping may be fascinating to combine *graphic methods* and *natural philosophy*.

Sometimes, it is found that literacy affects a lot of use of maps because it is most fundamental to appreciate graphicacy distinctively. According to Balchin, modern graphicacy, with its long history of development from simple sketch in the sand to a degree of sophistication has been recognised as a fundamental support, distinctive in kind and form of *macroscope* with reducing power to scale for perception. Such qualities, if generated continuously, the development of thematic cartography may get utmost attention in cause of socio-economic development.

Turning to the implications of statistics and thematic mapping, the scope, nature and form of graphicacy are essential to view specially in regard to composition of maps for resources. To make the job lucid, the entire gamut of the scope, nature and form is grouped into (a) Quantitative data, (b) statistical variables and transformation of maps' (c) spatial transformations of number and classes, (d) symbolisation and pictorial generation of statistical numbers, and (e) regional approaches for planning. Now let us, view these important variables and scan the system, methodologies and configuration as adopted in some of the important works in thematic mapping drawn from Indian and other countries literature on the subject for future research and common understanding of maps by the people and to meet the future challenges of the changing economic and geographic landscapes.

Quantitative Data

Numerical information mapping is one of the most important form for research generalisation both in geography/cartography and several social sciences specially demography, sociology, behavioural sciences and alike. It is well known that ratio scales to points, areas, geographical space generalisation, mapping national economic wealth such as GDP, GNP and income, matrices with multiple variables cartometric transfers of statistical realities etc. are suitably mapped with reference to administrative areas, national or in global terms. A lot of examples are available in the researches in geocartography for evaluation.

To mention a few the ratio scale with area and points are most common in economic studies of regions or nations such as population of cities graded in all classes in the maps of Indian censuses and the censuses of many developed and developing countries on the recommendations of the Population Commission of the I.G.U. through the decades (Olsson 1963). India also took a lead in preparing the population distribution on 1 : 1m by the NATMO under the editorship of late Dr. S.P. Chatterjee by coin system of mapping on the data of 1961 census (Dasgupta 1960). The national mapping of distributions of population on 1:2.5m for the entire country on the basis of 1981 census data by the Indian Census adding pressure of population in terms of density on the maps of the series has been carried out (Census of India, 1981). However, the combination of point ratios in terms of populations of cities and towns by hollow pies or three dimensional spherical or two dimensional square ratios to populations of cities and towns have been legitimate controversial issues through the decades. The cartographers take a liberty to reorient these point ratios of dimensions suiting to the requirement of scale of thematic maps. This matter, itself is a concern of great research in statistical thematic mapping according to scale and format of maps of regions and nations.

Quantified data, further, can be used in many ways on thematic maps and in this direction the *mental mapping* is yet another problem to tackle a perfect distributional layout. Gould (1974) has narrated several dimensions of such issues. In practice, besides many scholars, the work of Fuller and Chapman (1974) in testing the role of mental maps in migration research is quite impressive in highlighting the preferences in various components in the U.S.A. The scope of statistical derivation, since established earlier, mapping endeavours in Indian situations may be quite desirable in behavioural studies with mapping support on the issue.

Statistical Variables and Transformation of Thematic Maps

The challenge of cartographers in mapping statistical variables in transforming thematic maps is great and a lot of literature with novel innovations have been made in this direction. The simple procedures of mapping averages, mean of data sets, regressions and SD (standard deviations), ranksize and frequency distributions are of course used largely to support researches on the one hand for smaller areas and locations and finds a big gap largely in developing thematic mapping in view of resources mostly on account of dissimilarities of spatial coverage and relational inequalities in data. It is observed that regional maps are quite common with such derivations of statistics on maps, such as the well known learning exercises in *Elements of Cartography* (Robinson *et al.*, 1978). These are of course very useful and basic for students and practitioners of cartography. It is very much essential that cartographers may match the variable and reorient large scale maps. An example is cited with a problem of adoption in future years to the cartographers and cartographic institutions as below.

Let us conceive a problem to map out the occupational structure at district level for India on a base of 1:1m or 1:2m for example to establish a trend of inter district distributions of workers and its association with cultivators, agricultural labourers and non-workers in the sub-set as prudently selected by the cartographer. Now the theme being so simple in case for India but mapping and endeavour may pose a great challenge to establish associations by transfacing two sets of statistics (a) proportion of workers, and (b) establishing a tripartite relation with cultivators, agricultural labourers and non-workers in them. It is suggested that the variable on workers statistics may be layered in terms of statistical grades and the diagrammatic coherence by statistical tripartite diagrams (Roy, 1971) may be matched districtwise. If such a thematic map comes to light for the entire nation, on the scales proposed above, a great deal of experimentation and objective evaluation can be generated for a fruitful research. This can be also matched in time series. Alas! someone can come up to fulfill this task.

On other counts, it is mentioned that synthetic mapping on the thematic front can also be generated. This synthetic mapping has to inherit the concepts of stochastic process, score adjustment or of the qualities of retrogression on sound judgement. Works on methodologies are numerous in the western literature on these aspects but the cartographic models which lack substantially in Indian situations obviously are due to reasons related to lack of knowledge in statistics and mathematics in present day geocartographers because of gaps in their training.

This system of mapping has a long historical development. Taking the liberty to draw examples from earlier researches, the most characteristic example of a simple nature can be cited from the work of Smith (1969) who very ably brought out industrial trends by measuring structural changes in employment in 14 industrial categories in the northwest Britain and put forth six synchronous indexing measure by coefficients of diversification. Such maps are yet to come from the Indian cartographers and thematic mapping institutions in India on large scales by integrating associations of data sets.

It is strictly noted that synthetic aspects of cartographic assessment was, of course, carried out on an Indian problem on the urban-rural duality in the regional structure with an example of Andhra Pradesh by Berry & Rao (1968) on factor analysis with special reference to urbanisation and modernity, rural plurality, females in agriculture, certain characteristics of land use, Scheduled Castes and their relation of areal units (Taluks) by factor analysis. This research could not be strengthened by Indian cartographers probably due to lack of computers and computer training.

On another front to measure the well known physical quality of life index (PQLI) measure having unique potentiality of mapping could not emerge in India so far. Recently a series of methodologies on mapping population regions (Roy, 1982) (Fig. 4.1) and PQLI measure of development (Roy, 1985) were introduced in these realms of mapping systems. In addition, a national perspective to identify hierarchic land and planning nodes in India was first time projected in the national volume of the Census Atlas (1981) to create a consensus to match the nodes and non-agricultural environs for acceptance (Fig. 4.2). The results were quite encouraging because these two systems could project India's situation in terms of dichotomies and tracheotomies of population distribution on the one hand and problems of integration of basic resources with literacy, development and resource regionalisation by statistical tests. The results were quite convincing even to the layman, bureaucrats or politicians to take a judgement for development. Such researches may need to be strengthened in thematic mapping synthesising for statistical variables on current problems for a forceful tool of investigation. It is emphasised that spatial and schematic mapping models can also absorb these methodologies to project real situations of our economy and landscape for development by bands of cartographers in future.

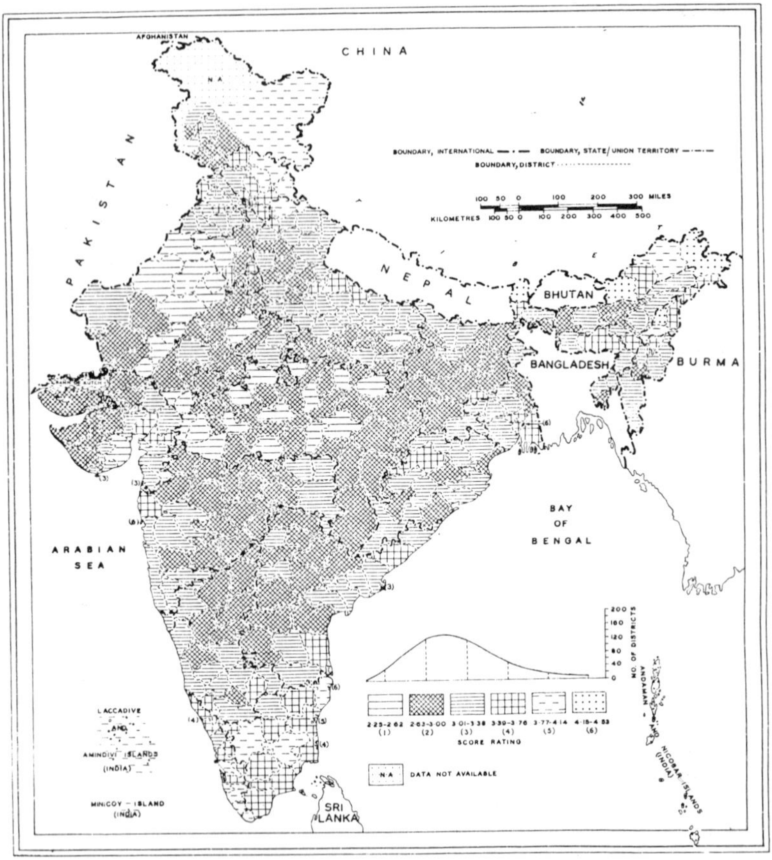

FIG. 4.1: India — Composite Demographic Regionals

Spatial Transformation of Numbers and Classes

The gamut of spatial transformation of numbers and classes is one of the most important and basic tools in geocartography and thematic mapping. The entire basis of making of geocartographic models lie on this aspect which is related to fixing of intervals of gradations of a parameter both in terms of number and smoothed data sets. Numerous maps and diagrams are enriched through the world and it is not a hard time to spot out any example at easy hand. However, the scientific layout of maps vary a lot at individual and institutional levels as well.

While opening an atlas or national atlas of any country, the bulk of the maps having relevance to statistical consumption of data for mapping, it is observed that the grading (class intervals), series intervals and data manipulations to fix classes are various. Sometimes, rather in most of the cases it is independent on maps of series of the same system of mapping. This analogy creates a gap in evaluating the maps for specific objectives. Sometimes it happens due to the ignorance of the cartographers and in cases it is unavoidable because of non-synchronous data sets and change in areas of administrative units.

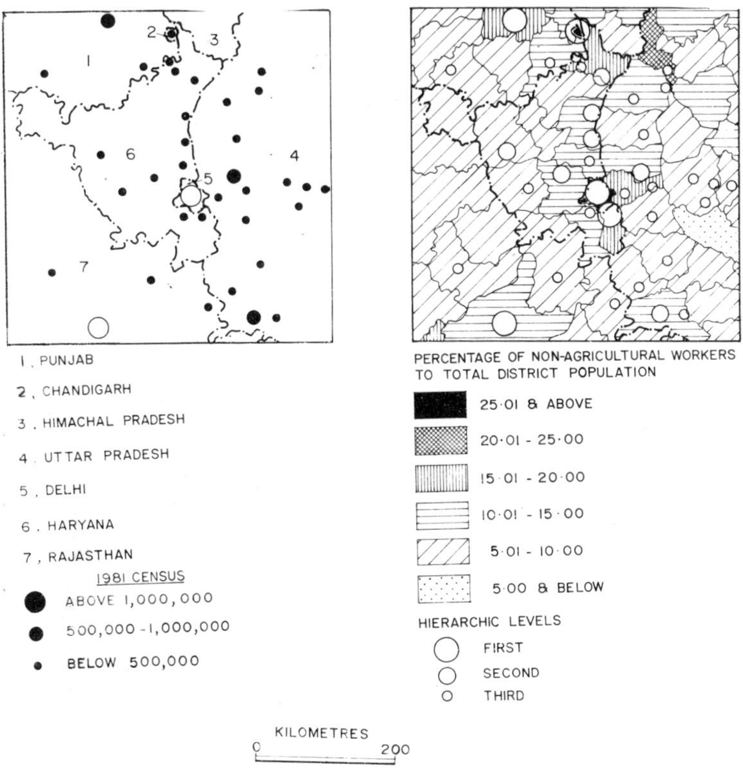

FIG. 4.2: Hierarchical Levels and Planning Nodes

It is underlined to modify the above statement also that cartographers may not be blamed for this anomaly but the class intervals are generated at times unprudently. The authors' experiences show that the statistical sets or variables often times (rather in most cases) change the definition of variables; sometimes the areal units are changed for administrative purpose and also data gaps create a mere problem to the cartographer to align uniform class intervals of a time series map on even a singular theme. The authors' own work in composing time series maps on literacy (1901-1971) in the *Indian Census Centenary Atlas* (1871-1971) in plate numbers 77 through 95 had discreet class intervals in total, male and female literacy due to gaps of data in historical perspectives of the population and areal changes. However, a logical attempt could solve the problem in only specific areas (districts) and maps were appreciated by the demographers and others for their further researches. This is one of the cases which can be readily cited to take a lesson for future caution for cartographers working on such problems in the country.

In addition, it is also revealed that even a computer (may be sophisticated) if the data input ability to manufacture in the form of isogramic models of various areas or locations of thematic mapping may do the same error unless the uniform data for uniform areal units are not fed in. This is probably one of the reasons that the cartographers don't embark upon for a time series maps even on isograms measuring trend surfaces, say, for example, the 3D representation of GDP per capita in South Africa (Smith, 1975) and G.F. Jenks statistical surface for population densities in civil divisions of Kansas could not be popular for series mapping for adoption in thematic atlases. However, there is no doubt that these models are quite forceful for bringing the problem of representation to the eye and the statistical valuation is difficult to interpret, rather it remains to give a notional picture of the art.

The other problems in relation to the group of issues on statistical information and classes of mapping, the maps related to choropleths, isopleths, chorochromatic and similar generalised map — techniques are common and understood by most people for perception and studies. It is, therefore, quite relevant that the cartography either in black and white or in colour should be so smoothed near to realities with the legend and classes thereon, that the users may actually find useful to the problem. Most of the statistical materials related to population such as growth, density, number, ratio to a theme variable may be drafted in such a manner that it may strike the user instantly and the work of the cartographers is respected. This is possible, in addition, by involving and inserting the data magnitudes in such a manner that the users get a dimension to the pattern mapped with areal units and numerical strength noted for immediate reference to the actual statistics (Fig. 4.3 and 4.4). This trend needs to be brought out in the gamut of mapping of data in the realm of thematic mapping which so far was avoided to limit congestion on maps. Thus, it is noted that the input on maps become a single phased mapping, i.e., the map 25 (or any other map) in the *Atlas of Kentucky* (Karan *et al.*, 1977) on Employment of tertiary activity since mapped in five classes in colour gives a characteristic geographical distribution but lack the statistical value and it is felt that the actual number of employment in this activity in rounded figures if would have been given, it would have served not only geographers but all concerned. Violations of such nature of traditional cartography should be considered by professionals. Such situations are needed because the maps may be used fruitfully by non-geographers to satisfy their lust to draw what they need from the maps. Indian cartographers may probably think to reshape the geographical cartography for the thematic maps on this consideration. The fear of inherent difficulties of space limitations and use of letterings and typography can be overcome by computerised composing of letterings for which so many options may be available in the post 1990 phase of development of hardwares and this may be satisfied.

Symbolisation and Pictorial Generalisation

The quality and its successful use of symbolisation and pictorial generalisation on thematic maps have been always significant but at the same time drab. A large number of national atlases adopt the pictorial oriented scales at about 5 per cent share of the total number of maps in an atlas. The *National Atlas of India* of NATMO and the national *Census Atlases* of the Census of India (1961, 1971 & 1981), have prudently discouraged generalisation by pictorial models of the statistical transformation.

Symbolisation and pictorial concepts in thematic mapping are precisely very vague but complex. Let us not equate now pictorial as a picture of a bag symbol equals to 110 metric tonnes. But symbolisation and pictorial concepts in cartography as is well known relate to various forms of cartograms ranging from dot methods, pie, proportional circles, divided circles, symbolising the matching of statistical data to a large number of representations from that of a dot map of population to the complex flow determinations of quantities of dimensions.

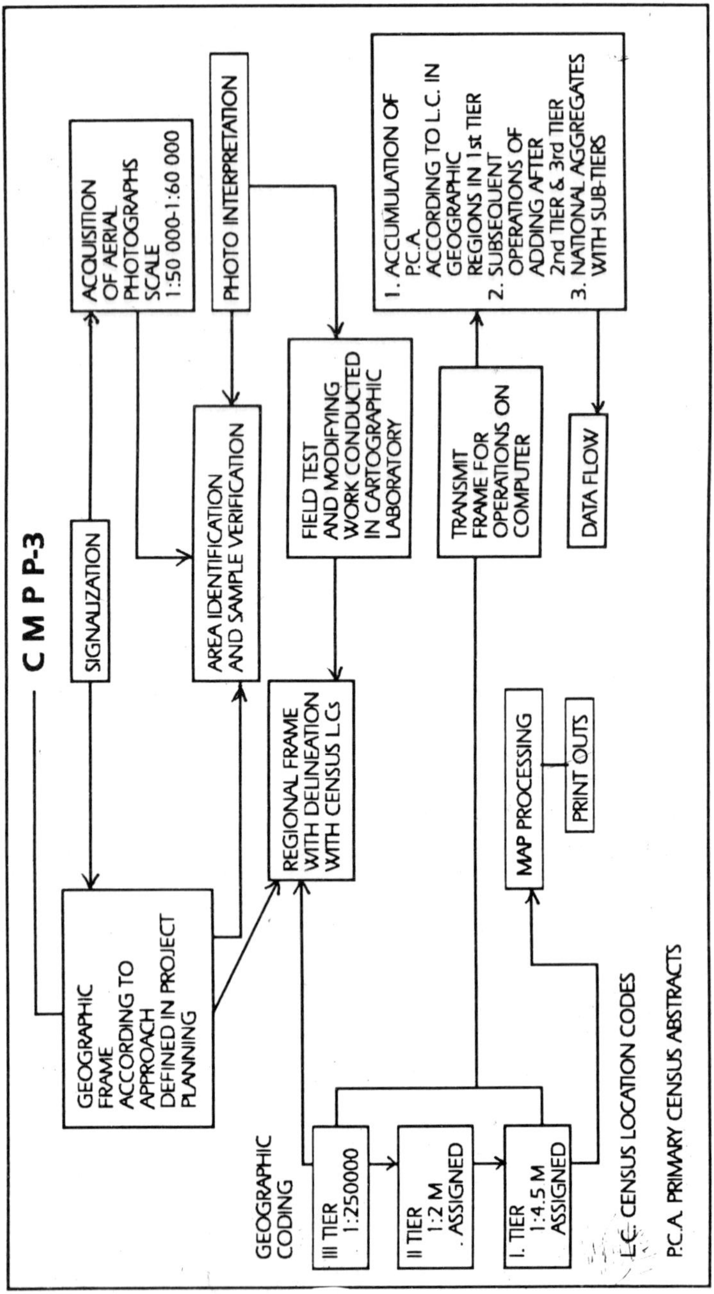

FIG. 4.3: Technical and Procedural Flow

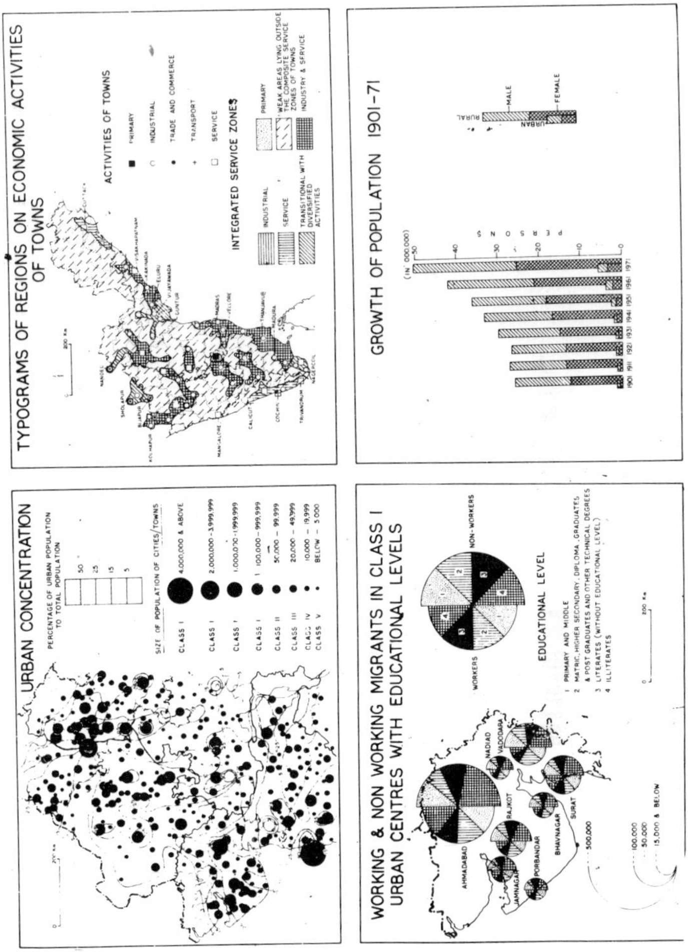

FIG. 4.4: Urban Concentration. Typograms. working and non-working migrants and Growth of Population

It, in itself, is a great task to deal with reference to this subject on statistics and thematic mapping. A large number of commonality lie in using such techniques by cartographers dealing with specific mapping. The choices are great and at the same time certain traditionality has been pervading in tackling mapping since long. The use of these techniques are usually well understood by the non-cartographers and appreciated as well. Since the scope of generalisation is most limited to such techniques, there is a good deal of correspondence of the forms selected by the cartographers on specific themes to the relevant statistics which generates generalisation automatically with the lowest and highest data and forms come out according to scales.

The recently brought out *Atlas of South Asia* (Dutt & Geib, 1987) brings out certain very good examples of symbolisation on maps on small scales. It drives clearly to the users to assess the problems and patterns of the statistics represented. In addition, the use of such techniques have been made in the census thematic mapping of India on 1:4.5m scale to a great measure earlier. A particular mention is made here of the 15 plates of maps created specially on cartography of migration first time (Roy, 1988). This success further prompted to bring out 20 plates of 26 maps integrating the 1971 and 1981 migration data as one of the programmes of the 1981 decade mapping plan. The mapping designs consumed a lot of time for continued research on the concept of *spatial* and *iconic* models in the mapping of India's internal migration (Roy, 1972-73) by combining a mass of statistics transformed models of range grading and scaling, leading to chorochromatic, statistics rationalising psychological to symbolise qualities of migrations, innovations in segment-ratio relations, and flow of migrations with combinations to view the balance of migration as accepted by demographers in the country as a step forward in providing tools for all concerned. The scope of enlarging the techniques are still conceived and hoped that the future professionals in the themating mapping may come forward to further develop the concepts of mapping in this branch and elaborate the techniques in future censuses.

Regional Approach in Cartography

Regional approaches in cartography may be three faced, viz., (a) non-statistical, (b) statistical oriented, and (c) cartometric. Among the non-statistical identification of regions, areas are classified based on derivations from large scale (in some cases small scale) maps mainly topographic, geologic and other similar backgrounds leading to the formation of homogeneous areal units for considering either physical regions, soil regions, or other subjective activity. In relation to such regions, the statistical sets are generated to highlight issues subjectively in accordance with the regions. The statistical oriented regional approach for cartographic evaluations are based on multiple criterion with statistical-resources baisness directed mostly to multipurpose regions for development is one of the forceful mapping tools. The Indian Census has taken the lead by producing 30 volumes recently (Census of India 1988-89) on this basis in its series on the Regional Divisions of India. The technical procedures to build such national exercises are defined in Fig. 4.5.

The cartometric as one of the approaches to regionalisation problems may be quite fascinating on the issues related to population and resources analysis of local and regional areas. The authors' work on *Urban Corridors* in India according to this system has been accepted as one of the tools in tackling the parametres of urbanisation problems by National Commission on Urbanisation 1988. Although the system for *Urban Corridors* are generated on small scale, a great deal of future work on large scale, if created by cartographic institutions, may serve as base line for regionalisation of such areas which the Union Government may be conceiving to bring about change in the set up which may lead to help local problems of urbanisation to authorities concerned. Such elaborations may be most useful in the present day context. There are several examples to satisfy specific objective, such as *Grid Mapping* leading to establish statistical areas for conduct of the census as adopted in Japan since 1965 and the U.K. Census of the 1970. The frames in these countries are also devoted to the extensive mapping of demographic characteristics to a greater degree of success as special areas for mapping.

Due to advent of computer and digital mapping, the field of grid mapping is expected to be enormous in future. A most characteristic example of establishing relationship in two dimensional mapping by taking educational attainment and per capita income variables in the U.S.A. showing colour choropleth in grids is yet another pitch in statistical mapping (Meyer, Broomee & Schweitzer, 1975). However, traditional cartography has played so far in a significant manner to establish grids and measures some parametre of statistical mapping on limited scale in India recently. The Census of India as an

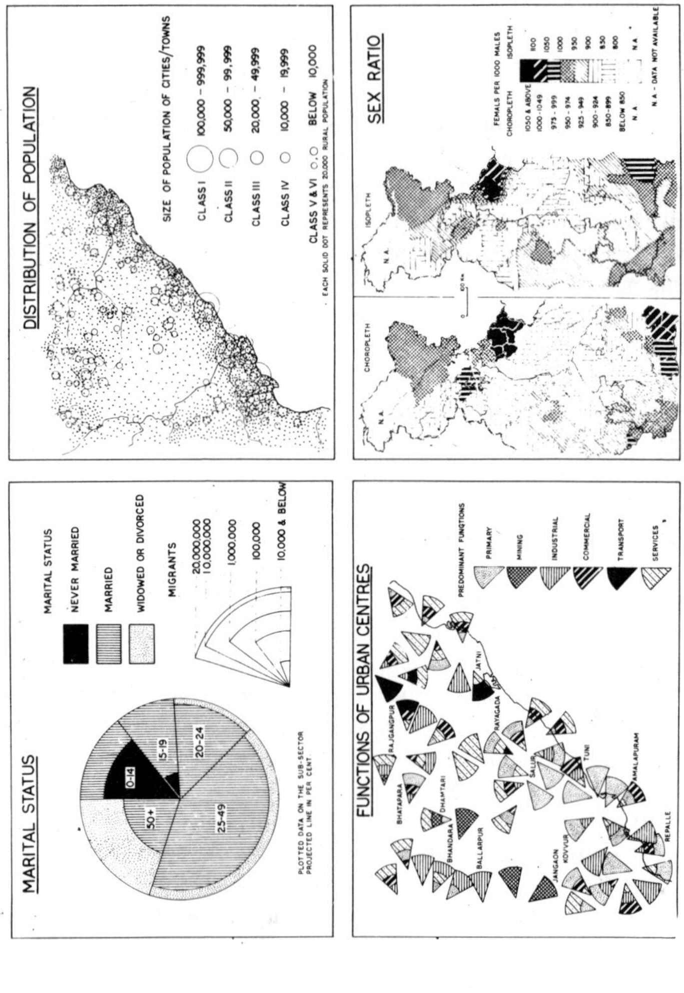

FIG. 4.5: Marital Status, Distribution of Population, Functions of Urban Centres and Sex Ratio

experimentation evolved 10,000 km^2 grid to evolve a new approach to urbanisation by placing the urban settlements of all classes and measured the urbanisation with reference to such grids for the entire country at 1:4.5 million scale as an experiment in the 1981 census Atlas of India. It is planned now to enlarge this attemp and introduce computer technology in the Map Division of the Census of India to initiate various researches including the initiation of Geocoding for the A.D. 2001 census of the country. A long toil and continuous thinking are essential to tie and replace the traditional cartography to digital. A massive plan is already placed for the consideration and establishment in this direction.

Concomitantly, the Indian Census took the lead to help the district level planning of the Eighth Plan of the country by giving an approach of regionalisation known as *Regional Divisions of India — A Cartographic Analysis*, by giving homogeneous regions from the lowest unit (sub-micro) of the district connecting to the micro, meso and macro regions of the nation in 30 volumes for immediate use of the planners and planning administrators for the country. This entire cartographic venture has produced a mass of statistics classifying rural, urban and a landscape of cumulation of villages as sub-micro units at 4th tier levels. This mapping device has brought further the Indian Census Organisation as a leading applied-cartographic institution in the world. It is also thought upon to produce a lot of data-sets on population for analysis in forthcoming censuses of India and ensures a speedy development in digital cartography for planning. This system may be tied up conveniently for grid-mapping when the automation plan of the Census of India is implemented.

Summary

This chapter could project certain views on the statistical thematic mapping to establish processes and patterns and their relations to the development of cartographic art. Although, it could focus on some selected issues to identify problems and patterns of cartographers dealing with the thematic mapping on the basis of statistics and other data set, yet scholars and practitioners of cartography may scrutinise these views as presented to develop the discipline more vigorously. Wherever literatures are cited in the text to the paper, it is to bring the essence of the methodologies to further streamline by consulting and to ponder over more to develop the professionalism in the field and here starts another beginning of the phase of challenges to make the art of cartography suitable to India's needs. However, the cartographers are not supposed to create maps only for technicians rather it should be also for the laymen and that is the success to grow in the profession and for the art.

REFERENCES

Balchin, W.G.V., (1970), *Geography: An Outline for the Intending Student*, Routledge, London.

Berry, B.J.L. & Rao, V.L.S. Prakasa Rao (1968), Urban-Rural Duality, *Geographische Zeitschrift*, Franz Steiner Verlag, Weisbaden.

Census of India (1961), *Census Atlas*, Series I (Pt I), New Delhi, (Direction P. Sengupta).

Census of India (1971), *Census Atlas*, National Volume, Series 1(i), New Delhi, (Direction B.K. Roy).

Census of India (1974), *Census Centenary Atlas*, 1871-1971, New Delhi, (Direction B.K. Roy).

Census of India (1981), *Census Atlas*, National Volume, Series 1(i), New Delhi, 1989, (Direction B.K. Roy).

Census of India (1988-89), *Regional Divisions of India — A Cartographic Analysis*, Series 1, Vols. I-XXX, New Delhi, (Direction B.K. Roy).

Chatterjee, S.P. (1949), *Bengal in Maps*, Orient Longman, Calcutta.

Dasgupta, S.P. (1960), 'The Coin System in Population Mapping', *Geographical Review of India*, Vol. 22(2).

Dutt, Ashok & Geib, M., (1987), *Atlas of South Asia*, Westview, Boulder.

Freeman, T.W. (1964), *A Hundred Years of Geography*, Aldine, Chicago.

Fuller, G. & Chapman, M., (1974), On the mental maps in migration research, *International Migration Review*, Vol. 8(4).

Gould, P., (1974), *Mental Maps*, Penguin.

Govt. of India (1988), Report on the National Commission on Urbanisation, Vol. 2 (Pt.I), New Delhi.

Jenks, G.F., (1967), The data model concept in statistical mapping; *International Yearbook of Cartography*, Vol. 7, Gutersloh.

Karan, P.P., & Cotton, M. (1977), *Atlas of Kentucky*, The University Press of Kentucky, U.S.A.

Meyer, M.A., Broome, F.R., & Schweitzer, R.H.(Jr.) (1975), Colour Statistical Mapping by the U.S. Bureau of the Census, *The American Cartographer*, Vol.2.

Robinson, A., Sale, R., & Morrison, J., (1978), *Elements in Cartography*, John Willey, New York.

The National Geographic Magazine, *Span*, (1989), Vol. XXX(6).

Roy, B.K. (1972-73), Cartography of migration maps, Bombay *Geog. Magazine*, Vol. 20-21(1).

Roy, B.K. (1982), Indian Census Cartography, *Geojournal*, Vol. 6(3).

Roy, B.K. (1982), Population Regions — A Perspective in Geographic Space of India, *Geojournal*, Vol. 6.2.

Roy, B.K. (1985), PQLI measure of development, *Geojournal*, Vol. 10.1.

Roy, B.K. (1987), The need for automation in India, *Geographical Review of India*, Vol. 49(2).

Smith, D.M. (1975), *Patterns in Human Geography*, Penguin, London.

William Olsson, W., (1963), Report of the I.G.U. Commission on a world population map, *Geographiska Annalar*, Vol. xlv(4).

5

GEOGRAPHICAL INFORMATION SYSTEM

SMITA SENGUPTA

The primary goal of geographical information systems is to take raw data and transform it, via overlays and analytical operations, into new information which can support the decision-making process. The word *Geographical* in GIS implies data with specific locations. In a geographical information system, the data should be referenced in a manner which will allow retrieval, analysis and display on spatial criteria. So, researchers and decision-makers compare and combine thematic maps, ground survey information and remotely sensed data of the same area on a proper geodetic grid. The reduction of many different *layers* of data into a new layer of information, usually in cartographic format is required in a decision-making process. The new information can consist of simple overlays to complex simulations.

This broad view of geographical information system encompasses many disciplines and applications. A geographical information system is not a computer based system by definition, but it is generally considered to be such. Although computer processing has been applied to geographical problems for more than a quarter of a century, it appears that there must be something different and more important about this subject that has caused so many diverse group to take notice and invest so many resources on implementing new systems.

Evolution of GIS and Computer Cartography

The collection of data about the spatial distribution of significant properties of the earth's surface has long been an important part of the activities of organized societies. From the earliest civilizations to modern times, spatial data have been collected by navigators, geographers and surveyors and rendered into visual form by map makers or cartographers. Originally maps were used to describe far off places as an aid for navigation and for military strategists.

The roots of geographic information system data back to the mid-eighteenth century with the development of cartography and the production of the first accurate base maps. Thematic mapping, that is the definition of attributes to a given geographic area, came about soon after, with maps portraying magnetic variation with isolines and wind direction by means or arrows. The idea of recording different layers of data on a series of similar base maps was on established cartographic convention by the time of the American revolutionary war.

Advances in the physical and social sciences provided geographers with the tools for spatial analysis. Statistical techniques, number theory and advanced mathematics were all developed. By 1835 technology (advanced cartographic techniques), science (social science theory) and social thought (environmental responsibility) had advanced to the point that the combination of these three factors had supported more comprehensive thematic mapping projects. But the industrial revolution was the main catalyst in the evolution of GIS. The explosion in manufacturing increased the need for raw resources, drew people into crowded environments, and created the need for an extensive infrastructure, both social and industrial, especially in the transportation field. The Atlas to accompany 'The Second Report of the Irish Railway Commissioners' (1838) was consisted of a series of maps that depicted population, traffic flows, geology and topography and was perhaps the first Geographic Information Systems per se.

Thematic mapping continued to improve with the London publication in 1848 of the first truly uniform world atlas in respect of scales, symbolization and cohesiveness. Such convention as isobars, graduated circles and volume lines were used extensively. Thematic maps continued to developed in scope and purpose. The years between 1835-1959 are the 'Golden Age' of cartography. During these years many techniques were innovated.

The use of electronic computer in the late 1940's for electronic calculation opened new possibilities of research based on the massive manipulation of huge data file. This has also helped to generate intricate models and alternatives. In the late 50's meteorologists, geophysicists and geologists have incorporated computer generated maps into their works. However, crude, at least the capability to print out geographic data in a cartographic format had been established.

During the 1960's and 1970's, new trend arose in which maps are used for resource assessment, land evaluation and planning. The people started evaluating the different aspects of the earth's surface in an integrated, multidisciplinary way, using two methods. The first attempts to find ***naturally occurring*** environmental units that can be recognized, described and mapped in terms of the total interaction to the attributes under study was termed as *gestalte* method by Hopkins (1977). According to Vink (1981) there must be a recognizable, unique and interdependent combination of the environmental characteristics of land form, geology, soil vegetation and water. He termed it as *correlative complex.*

The main problem with the results of integrated surveys is that for many purposes they are too general and it is often very difficult to get specific information from them about particular attributes of landscape. So there has remained a ready market for the more conventional monodisciplinary surveys, such as those of geology, landform, soil, vegetation and landuse, particularly as surveys have tended towards larger and larger scale as development has proceeded.

The availability of wide range monodisciplinary research maps has made users to seek for a combined and integrated overview of these available information. In the U.S. planners and landscape architects started combining and integrating several monodisciplinary resources simply by overlying transparent copies of the resource maps and looking for the places where the boundary on the several maps coincide. One of the best known exponents of this simple technique was the American landscape architect, Ian McHarg. In 1963, H.T. Fisher started using computer to make simple maps by printing statistical values on a grid of plain paper. Fisher's program SYMAP, short for synagraphic mapping system, includes a set of mudules for analysing data, manipulating them to produce choropleth or isoline interpolations, and allows the results to be displayed in many ways using the overprinting of lineprinter characters to produce suitable grey scales.

There are other programmes like GRID and IMGRID that allowed the user to do in the computer what Mcttarg had done with transparent overlays. All these programmes were similar to that of the work of Mcttarg only they allow the work to be done more quickly and reproducibly. Gradually, the users found is also possible along with simple overlay mainly in planning and ecological studies. SYMAP, GRID, IMGRID, GEOMAP and many of these relatively simple programmes were designed mainly for quick map analysis of grided data. Initially their results were displayed only by using crude lineprinter graphics but many cartographers refused to accept the results they produced as maps. Cartographers had began to adopt computer techniques in the 1960's, but these were until recently largely limited to aids for the automated drafting and preparation of masters for printed maps.

By the late 1970's there had been considerable investments in the development and application of computer assisted cartography, particularly in North America by the government and private agencies. Hundreds of computer programmes and systems were developed for various mapping applications. There are estimated to be approximately more than 1,000 geographical information systems now installed in North America. In Europe the developments proceeded on a smaller scale than in North America but major strides in using and developing computer assisted cartography have been made by several other nations, notably Sweden, Norway, Denmark, France. The Netherlands, United Kingdom and West Germany.

The history of using computers for mapping and spatial analysis shows that there have been parallel developments in automated data capture, data analysis and presentation in the several broadly related fields. These fields are cadastral and topographical mapping, thematic cartography, civil engineering, geography, mathematical studies of spatial variation, soil science, surveying and photogrammetry, rural and urban planning, utility networks and remote sensing and image analysis. Military applications have overlapped and even dominated several of these monodisciplinary fields. Consequently there has been much duplication of effort and a multiplication of discipline specific jargon for different applications in different lands. This multiplicity of effort in several initially separated but closely related fields in now resulting in the possibility of linking many kinds of spatial data processing together into truely general purpose geographical information systems, as technical and conceptual problems are overcome. Essentially, all these disciplines are attempting the same sort or operation — namely to develop a powerful set of tools for collecting, storing, retrieving at will, transforming and displaying spatial data from the real world for a particular set of purposes. This set of tools constitutes a *Geographical Information System.*

Data Structure for GIS

Geographical data are complicated by the fact that they must include information about position, possible topological connections and attributes of the objects recorded. The topological and spatial aspects of geographical data processing are different from other modern data processing system. Geographical data are referenced to locations on the earth's surface by using a standard system of coordinates. The coordinate system may be purely local, as in the case of a study of a limited area, or it may be that of a national grid or an internationally accepted projection.

All geographical data can be reduced to three basic topological concepts — the point, the line and the area. So, a map is a set of point, lines and areas that are defined both by their location in space with reference to a coordinate system and by their non-spatial attributes. A map is usually represented in two dimensions but there is no reason to exclude higher dimensions except through the difficulty of portraying them on a flat piece of paper. The map legend is the key linking the non-spatial attributes to the spatial entities. Non-spatial attributes may be indicated visually by colours, symbols or shading, the meaning of which is defined in the legend. For geographical information systems, non-spatial attributes need to be coded in a form which they can be used for data analysis. A region is a set of pixels, areas or polygons that are described by a single legend unit.

The main problem of computer is that it can not accept geographical data directly. The data must be programmed to represent phenomenological structures appropriately. Moreover, the way of the geographical data are visualized by the user is frequently not the most efficient way to structure the computer data base. All these data have to be stored in magnetic devices that need to be addressed in a specific way. It is easy for the man to recognize shape and form of an object but in case of a computer it should be instructed to handle and display spatial data. The spatial data can be entered in the computer in two contrasting but complimentary ways. These systems are generally referred to as explicit or raster and implicit or raster and implicit or vector ways of describing spatial entities. The explicit representation makes use of a set of points on a grid while implicit representation means the use of a set of lines defined by starting and end points and some form of connectivity. (Fig. 5.1)

The simplest raster data structures consist of on array of grid cells or pixels. Each grid cell is referenced by a row and column number and it contains a number representing the type or value of the attribute being mapped. In raster structures a point is represented by a single grid cell, a line by a number of neighbouring cells stung out in a given direction and an area by an agglomeration of neighbouring cells. This type of data structure is easy to handle in the computer, particularly with programming languages such as FORTRAN. In the raster system each cell is in a two dimensional array so can only hold one number. The different geographical attributes must be represented sets of cartesian arrays, known as *overlays*. The overlay idea for separating data is not restricted to computer cartography. This process having been used by cartographers for preparing printed maps and by landscape planners or a very long time.

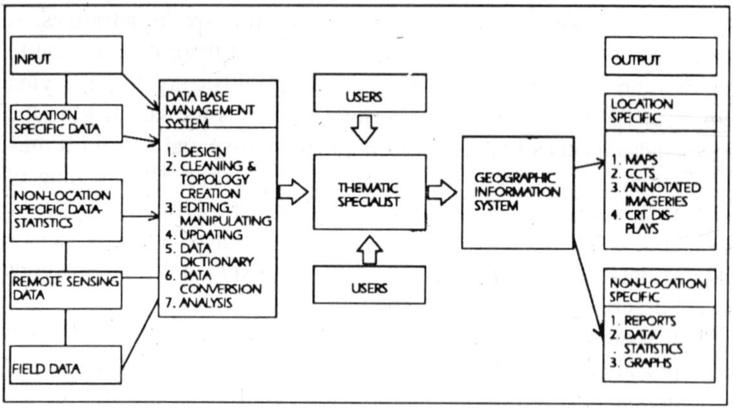

FIG. 5.1: Structure of GIS

In the vector system, an object is represented as exactly as possible. The coordinate space is assumed to be continuous, not quantized as with the raster space, allowing all positions, lengths and dimensions to be defined precisely. Point entities can be considered to embrace all geographical and graphical entities as positioned by a single coordinate pair. Line entities can be defined as all linear features built up of straight line segments made up of two or more coordinates. Areas of polygons or regions can be represented in various ways in a vector database. Most kinds of thematic mapping used in geographical information systems have to do with polygons. The aim of a polygon data structure is to be able to described the topological properties of areas in such a way that the associated properties of these basic spatial building blocks can be displayed and manipulated as thematic map data (Fig. 5.2).

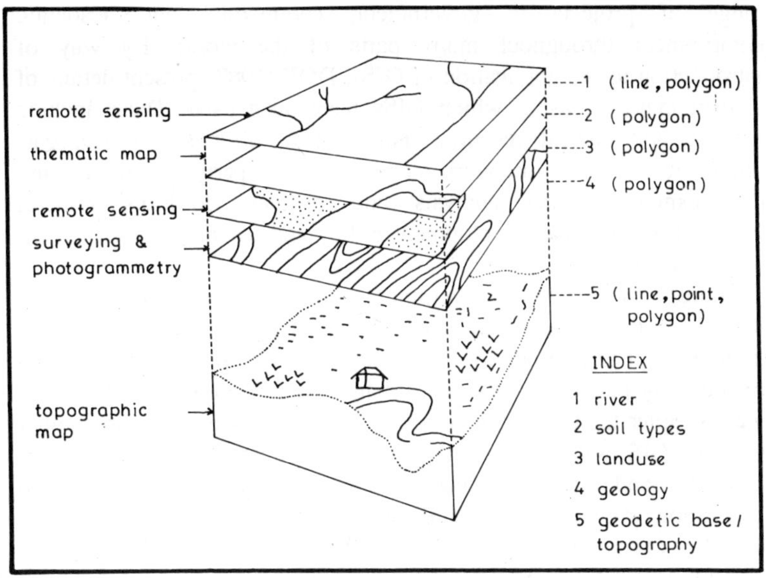

FIG. 5.2: Geographical Information System

The raster and vector methods have distinctly different approaches for modelling geographical information. The user was faced with the choice of raster methods that allowed easy spatial analysis but resulted in ugly maps or vector methods that could provide database of manageable size and elegant graphics but in which spatial analysis was extremely difficult. The problem of raster or vector disappears once it is realized that both are valid methods for representing spatial data, and that both structures are interconvertible. Conversion from vector to raster is the simplest but the reverse operation is complicated because it requires the reduction of the numbers of coordinates in lines by a process known as weeding. In some circumstances it may be advantageous to have spatial data present in both raster and vector form particularly when line or boundary data need to be represented by connected network or drawn in a particular style and spaces between must be filled with a print raster of a given symbol or colour.

Application of GIS

The applications of GISs are legion. Already they are used for a vast range of projects in government, commercial and academic environment throughout many parts of the world. By way of introduction to the application of GISs, DOE (1987) present details of sixteen major areas in which GISs have been used. These include landuse and natural resource management, utilities (electric, gas, telephone and water) network management, property registration and development, market analysis and business location, mineral exploitation, regional development policy and teaching. Few of them are discussed below.

Local and central government bodies in several countries throughout the world are using GIS for organising and integrating data in an attempt to improve the efficiency of service provision (Fig. 5.1). In the early part of 1970's new urban models were developed which were simpler than the original ones. With the consistent improvement of new GIS with increasing capabilities, many of the long term inventory and mapping needs have been met. One of the second generation models that was developed in Australia in 1970 was TOPAZ (Technique for the Optimum Placement of Activities in Zones). TOPAZ is a general planning technique that has been applied at the regional, urban and facility planning levels. The model has been used in conjunction with urban geobased data to analyse sewer and water expansion and costs, trip generation and model choice, water catchment planning, investigating issues, originating with corridor developments and new communities and planning of future urban services. As the system can allocate activities to spatial zones. It has also been used to lay out buildings and complexes as a micro scale GIS. With the new TIGER system (U.S. Bureau of Census) and special commercial data files like that of ETAK (currently being coded for most metropolitan areas of the US) sophisticated application in navigation, vehicle routing, scheduling, dispatching and automatic vehicle location will become common place. The use of GIS in cities will expand beyond property records and planning functions of many real time applications as well.

The application of GIS in natural resource and environmental management has developed mainly in North America. In 1982, thirty-two states in the USA possessed statewide natural resource in formation systems. This systems has created a statewide 40 acre grid cell (raster) mapping system containing a range of natural resource data. This system has been used for projects system has been used for projects such as — an environmental assessment of the possible impact of mining, a study of seasonal homes and recreational activity, high way and powerline corridor routeing, solid, hazardous and radio active waste disposal setting, and modelling the vulnerability of equatic and terrestrial areas to acid precipitation. Topographic data are entered from the United States Geological Survey 1: 24,000 topographic maps. These are combined with thematic information about soils, geology, landuse/land cover, public ownership, hydrology, transportation and political subdivisions, some of which are derived from remote sensing satellite data.

Conclusion

The geographical information system is a tremendously exciting area of geography to be involved in at present. The rate of technological development and the speed at which research is progressing ensure that new ideas and techniques are being presented almost daily. The integrative and applied nature of GISs makes them very attractive to many geographers. They may help to bring together many disparate branches of the discipline and may provide a common focus perhaps once again based upon the region as the fundamental geographical concept.

REFERENCES

American Society of Photogrammetry (1960), *Manual of Photo Interpretation*, Washington DC.

Allam, M.N. (1983), Digital Mapping in disturbed system environment, *Proceedings of Auto-Carto 5,* ASCM, USA.

Anderson, D. *et al*. (1983), Application of a GIS — Identification of Resources Sensitive to Acid Deposition, *Proceedings of Auto-Carto 6*, ACSM, USA.

Baveja, S.D. (1987), *Geographic Information System — An Overview*, Digital Mapping Centre, Survey of India, Dehradun.

Callahan, G.M. (19830, Digital cartographic data production using raster scanning techniques, *Proceeding of Auto-Carto 6*, ASCM, USA.

Chrisman, N. (1987), Fundamental principles of geographic information systems, *Proceedings of Auto-Carto 8*, ACSM, USA.

Collins, S. *et. al.* (1983), Advances in GIS, *Proceedings of Auto-Carto 6*, ACSM, USA.

Davis, L.S. (1980), *Representation and Recognition of Cartographic Data*, Academic Press Inc., London.

Dangermond, J. and Harrison, J. (1987), *Urban Geographic Information Systems: The San Diego Design Experience,* Environmental System Research Institute, California, USA.

Environment Canada (1981), *Canada Geographic Information System — Overview*, Report No. R001000, Environment Canada.

Erik de Man, W.H. (1987), Planning a geo-information system — A process of strategic choice, *Proceedings of International Workshop on GIS*, China.

Goodenough, D.G. *et. al.* (1983), On the transfer of remote sensing classifications into polygon geocoded data bases in Canada, *Proceedings of Auto-Carto 6*, ACSM, USA.

Mac Donald, C.L. and Crain, I.K. (1985), Applied computer graphics in a geographic information system — problems and success, *Computer Graphics and Applications*, Vol. 5, No. 10.

Polker, T.V. and Crain, I.K. (1984), *Geographic Information Systems*, Environment Canada.

Shelton, R.L. (1981), Remote sensing and GIS—An unrealised potential, *Geoprocessing*, Vol. 1, No. 4.

Zobrist, A.L. and Nagy, G. (1981), Pictorial information processing of LANDSAT data for geographic analysis, *LEEE Computer*, Vol. 14, No. 11.

6

TECHNOLOGICAL IMPACT ON GEOGRAPHY AND THEMATIC MAPPING

J.R. SHARMA AND NARVADA SHARMA

Geography being a dynamic subject, it has undergone rapid changes at different times at different places as per the need and so is the case with thematic mapping with little less intensity. In last 15-20 years geography discipline has become widely popular with great importance not only it being 'mother of all sciences' but because of the technological impacts it has undergone, resulting in more accuracy, precision, depth and scientific logic. These technological impacts can be attributed to five techniques or factors which are not mutually exclusive, namely quantitative analysis, application of computers, automated computer cartography, remote sensing and image analysis and geographical information system.

Quantitative analysis

During 1950's and early 1960's geography went under important philosophical and methodological changes which have subsequently been termed as the quantitative revolution. This change made the discipline, from an essentially qualitative and descriptive subject into one which became increasingly concerned with the development of generalised laws and theories about spatial patterns using mathematical and statistical methods.

The quantitative revolution in modern geography has now made it possible to measure phenomena with greater precision and accuracy, to develop explanatory mechanisms of greater sophistication so as to further improve the accuracy of qualitative judgements and to uncover or unfold deeper layers of meaning of objective phenomenon, which less sophisticated tools were unable to achieve, and to facilitate the process of abstraction from empirical facts so as to generate concepts and theories. Raza has rightly mentioned in the introduction of the book entitled *Statistical Methods in Geographical Studies* by Aslam Mahmood, "the quantitative analysis has made possible to select quantitative indicators of quality in order to make qualitative assessments. Geographers move back and forth from qualitative assessments to quantitative measurements. These both are necessary components of research strategy. Qualitative evaluations provide purpose and direction to quantification and gets back from it the accuracy, depth and precision."

Quantitative techniques can be used for many purposes e.g. how much data should be collected so that an accurate understanding of phenomena can be developed. This is a type of cost-benefit analysis, where the extra benefit gained from collecting more data is offset against the extra cost involved in collecting them. Statistics can be used to describe, classify and summarize data, thus a complex collection of data can be reduced to a few numbers which represent the key elements. Using statistics it is possible to place error limit on data and finally statistics enable us to be more rigorous in our work and to avoid self-deception.

If quantitative techniques with the help of computers are seen, the speed and reliability of computers has allowed more complex statistical tests to be carried out on data sets of a larger and therefore, more realistic size. As a result, more time has been made available to do experiment with statistics. Perhaps most important, as far as geographers are concerned, is the fact that using computers for statistical analysis it is possible to place more emphasis on the interpretation rather than the calculation of statistics.

Use of Computers

Twenty years after the quantitative revolution geography once more underwent important change. The introduction of relatively low cost micro computers in late 1970's and 1980's and their pervasion into geography initiated computer revolution in geography. They have become so widely used because of great improvement in efficiency and effectiveness in a very short time. David J. Maguire rightly said in his book *Computers in Geography* that "in theory, there is nothing that a geographer can do with a computer that cannot be achieved by other means. In practice, however, there are many operations which would not be contemplated without the aid of a computer, because they are too time consuming. It is the ability of computers to perform simple repetitive operations very quickly and to store vast quantities of data that has made them so useful. Any application which can make use of these facilities can invariably be greatly enhanced when computerised." The development of statistical methods in geography has been closely linked with the improvements of computers. The ability to store large quantities of data which can be manipulated quickly in different ways, has made the use of computers almost obligatory for any type of complex statistical analysis. Indeed, prior to the advent of computers, some statistical analyses were simply too time consuming and complex to contemplate.

Computers are defined as simple electronic machines capable of the input, storage, manipulation and output of data. They are used to collect information about the environment, store information about people, manipulate statistical data, data management, computer cartography, remote sensing and image analysis, simulation and geographical information system. Today very powerful computers can work at speed of 1000 million instructions per second and can store 32 million words in the main memory. Linked to computers, graphic plotters can draw lines at one metre per second and line printers can print 500 characters per second.

Automated Computer Cartography

Cartography too has undergone rapid changes in last few years with the advancement in technology namely sophisticated equipments and use of computers in handling spatial data sets. With these innovations, the simple term cartography is better referred these days as computer aided or computer assisted or automated cartography. It not only includes maps and maps like products but also graphic images, terrain models and statistical graphs. Computer cartography, because of easy handling by using computers and more convincing nature, it is not only used in geography alone but in many subjects including widespread use in media.

David J. Maguire mentions that various methods which people use to develop, analyse and communicate ideas have been described on the four acies as literacy (written language), numeracy (mathematic symbols), oracy (spoken world – sometimes called articulacy) and graphicacy (cartographic and graphic images). Further he describes graphicacy as the communication of spatial information, this cannot be conveyed adequately by written or numerical means, and hence graphicacy (cartography) is one of the things which sets geography apart and tall from sciences and social sciences.

Quoting Rhind (1977), Maguire (1989) highlights the advantages of computer cartography, "It allows existing maps to be produced more cheaply and quickly. The map compilation time is much reduced, storage space is decreased and more importantly, the cost of updating and redrafting maps held in digital format is considerably less." Computer cartography facilitates the production of maps which are more closely tailored to user needs. Using a computer cartography system it is relatively simple to produce a map of selected features (for example roads, railways and rivers), for a selected area (the concept of sheet edge does not apply for digital mapping), at any required scale. The advances in computer cartography has led to greater experimentation in the mapping process. This is particularly the case where the maps depict statistical information. Users can easily try out different class boundaries, shading schemes and data transformations. Computers also allow the creation of new types of cartographic images, not previously possible using traditional means, e.g. three dimensional perspective plots and stereo maps pairs.

The thematic maps and topographic maps produced using computer cartography uses same principles, i.e. capturing digital map data called digitization using coordinates in vector or raster format, having single theme in each file for storage and output in the form of thematic or topographic maps with overlay of all files or of some files as per the requirement.

Remote Sensing and Image Analysis

Remote sensing defined by Colwell in the *Manual of Remote Sensing* is "the gathering and processing of information about the earth's environment, particularly its natural and cultural resources, through the use of photographic and related data acquired from an aircraft or satellite." The fields of remote sensing and image analysis have expanded rapidly in last 15 years and has become major thrust in geography and thematic mapping. Remote sensing devices are responsible for the collection of massive amount of environmental data at a regular interval and are routinely processed using specialist *turn key* image analysis systems which combine *state of the art* computer hardware and software. The vast scope for applications of remote sensing data has made it popular and attracted geographers for possible wider uses.

Remote sensing data collected in different spectral bands of EMR are analysed and displayed on both continuous and discrete images. In continuous images like aerial photographs having very high ground resolution, the data are in analogue format. In discrete images, such as the majority of those obtained from satellites, the data are in digital formats.

Remote sensing data has to go for great amount of transformation before they are used by sophisticated computer hardware and software. Various preprocessing techniques are applied to restore the images. Enhancement techniques to increase the information for different themes and extraction techniques for extracting the required information from the raw data. Major geographical applications where remote sensing data used are — detection, identification and mapping surface, subsurface and structural features; process of dynamic–volcano eruption, snow and ice, sea ice and iceberg production, clouds, rainfall etc., to classify and map plant, and soil distribution, the status of forestry and agricultural resources, production and yield, surface drainage, water, impact of irrigation schemes, desertification and change detection, landuse study, urban analysis, drought monitoring, salinity and water logging and finally preparing topographic and thematic maps.

Geographical Information System

It is indeed last 5-7 years that geographer have started using the real benefits to be derived from integrating data and programmes into large comprehensive systems, thereby opening up the prospects of many different operations and potential applications. Such integrated systems for the collection, storage, manipulation and presentation of geographical data are referred to as Geographical Information System (GIS).

Burrough (1985) and Goodchild (1988) define GIS respectively as follows:

"GISs are a powerful set of tools for collecting, storing, retrieving at will, transforming and displaying spatial data from the real world for a particular set of purposes".

"A GIS is best defined as a system which uses a spatial data base to provide answers to queries of a geographical nature".

Today there exists a tremendous amount of cartographic and thematic information for integrated and comprehensive use and produced very frequently by satellites and other sources and hence system like GIS has become essential for management of data sets and to use them.

All data that can be mapped has both locational (X, Y) and nonlocational (attribute) characteristics. For example, a feature may exist at X, Y location and possess Z attributes. These attributes can be both qualitative (landuse at a location) or quantitative (elevation at the same location). In addition the attributes at the location, attribute and time represent the contents of most GISs.

GISs offer to geographers new tools which significantly help to improve an understanding of spatial pattern and processes because of their quick and easy access to large volumes of data and to manipulate them: to select data by area or theme, to search for particular characteristics or features in areas, to update data sets quickly and cheaply, and to model data and assess alternatives. They are also used to crate new and varied types of outputs in the form of maps, graphs, address lists and summary statistics — all as per individual's needs.

Conclusion

These new techniques in geography and thematic mapping have made the subject very scientific and technical and the demand of geographers and cartographers have increased world over as scientists and engineers.

REFERENCES

Jensen, J.R. (1986), *Digital Image Processing*, Prantice-Hall, New Jersy.

Johnston, R.J. (1980), *Multivariate Statistical Analysis in Geography*, Longman, New York.

Maguire, David (1989), *Computers in Geography*, Longman, New York.

Mahmood, Aslam (1977), *Statistical Methods in Geographical Studies*, Rajesh Publications, New Delhi.

Parker, H. Dennisen (1988), The unique qualities of a geographic information system : A commentary, *Photogrammetric Engineering and Remote Sensing*, Vol. 54, No. 11, pp. 1547-49.

Peter, F. Fisher and Richard E. Lindenberg (1989), On distinctions among cartography, remote sensing and geographical information systems, *Photogrammetric Engineering and Remote Sensing*, Vol. 55, No. 10, pp. 1431-4.

Peter, L. Croswell and Stephen R. Clark (1988), Trends in automated mapping and geographical information system hardware, *Photogrammetric Engineering and Remote Sensing*, Vol. 54, No. 11, pp. 1571-6.

R.R.S.S.C. (1988), Lecture Notes on Digital Image Processing, Indian Space Research Organisation, Jodhpur.

Sahib, S. Ummer (1988), Education in computer cartography and geographical information system : An urgent need in Indian universities, *Annals of the National Association of Geographers of India*, Vol. VIII, No. 1, pp. 51-6.

Sharma, J.R. (1987), NNRMS-RRSSC and resource geography, *Annals of Associations of Rajasthan Geographers,* vol, VII, pp. 55-7.

Zhou, Qiming (1989), A method for integrating remote sensing and geographical information systems, *Photogrammetric Engineering and Remote Sensing*, vol. 55, No. 11, pp. 591-6.

7

NNRMS ACTIVITIES AND THEMATIC MAPPING

M.G. CHANDRASEKHAR AND B. MANIKIAM

The national seminar held at NRSA, Hyderabad in May 1983 took stock of the remote sensing applications in the country to various resources and themes, including ocean resources. A large number of end-to-end experiments carried out in collaboration with various user agencies were discussed and a set of recommendations were finalised at the seminar. One of the major recommendation was to set up a NNRMS with appropriate integration of conventional and remote sensing based information. The seminar concluded that remote sensing through airborne and satellite borne sensors covers surveying as well as monitoring, which are essential for the planning and management of national natural resources. Remotely sensed data has relevance in major sectors of the economy, such as agriculture, forestry, irrigation, human settlements, geology, ecology and oceans. Competence has been achieved in the use of this technique in areas like landuse mapping, evaluation of the ground water potential, regional geological mapping and substantial potential exists for its application to flooded area mapping, broad forest type mapping, estimating snow melt runoff and other numerous areas. Major national forest mapping and wasteland mapping have been done using remote sensing techniques. These techniques provides opportunity for speedy updating of maps, thus help monitoring afforestation efforts etc.

Based on the experience gained in the country in operation applications as well as noting many other potential applications, government has decided to establish the National Natural Resources Management System (NNRMS) which will integrate the data obtained through remote sensing into the existing system with appropriate technical, managerial and organisational linkages (Fig. 7.1). The scope of NNRMS is to offer an efficient, integrated, cost-effective and timely information system that enables judicious and planned utilisation of the natural resources of the country with due attention to the environment. All agencies connected with national development are being energised to utilise remote sensing data as a major new tool with its manifold applications. Remote sensing application experiments at district, state and regional levels are planned for the enhancement of competence and for widening their use.

As a follow up of the seminar, Planning Commission set up 8 task forces covering various natural resources to study and recommend the possible applications of remote sensing. The Task Force on Oceanography, Marine Resources and Coastal Studies headed by then Director, NIO gave a number of possible areas of remote sensing applications and recommended the setting up of appropriate facilities. Presently the Standing Committees on various themes headed by respective Secretaries at the central level are overseeing the progress of the various activities under NNRMS.

Space and Ground Segment

In order to ensure necessary remotely sensed data, Department of Space (DOS) has set up the Earth Station at NRSA, Hyderabad which is capable of receiving data from various satellites, such as LANDSAT, SPOT, IRS-1A and NOAA. The space segment consist of the IRS series of satellite, the first of which, IRS-1A was successfully launched during March 1988 and is providing regular data. It is planned to have a microwave remote sensing programme to meet the requirements of ocean resources management and certain selected land related applications.

Remote sensing through airborne and satellite borne sensors covers surveying as well as monitoring, which are essential for the planning and management of national natural resources. Remotely sensed data has relevance in major sectors of the economy, such as agriculture, forestry, irrigation, human settlements, geology, ecology and oceans. Competence has been achieved in the use of this technique in areas like land use mapping, evaluation of ground water potential, regional geological mapping and substantial potential exists for its application to flooded area mapping, broad forest type mapping, estimating snow melt runoff and other numerous areas. Major national forest mapping and wasteland mapping have been done using remote sensing techniques. These techniques provide opportunity for speedy updating of maps, thus help in monitoring afforestation efforts etc.

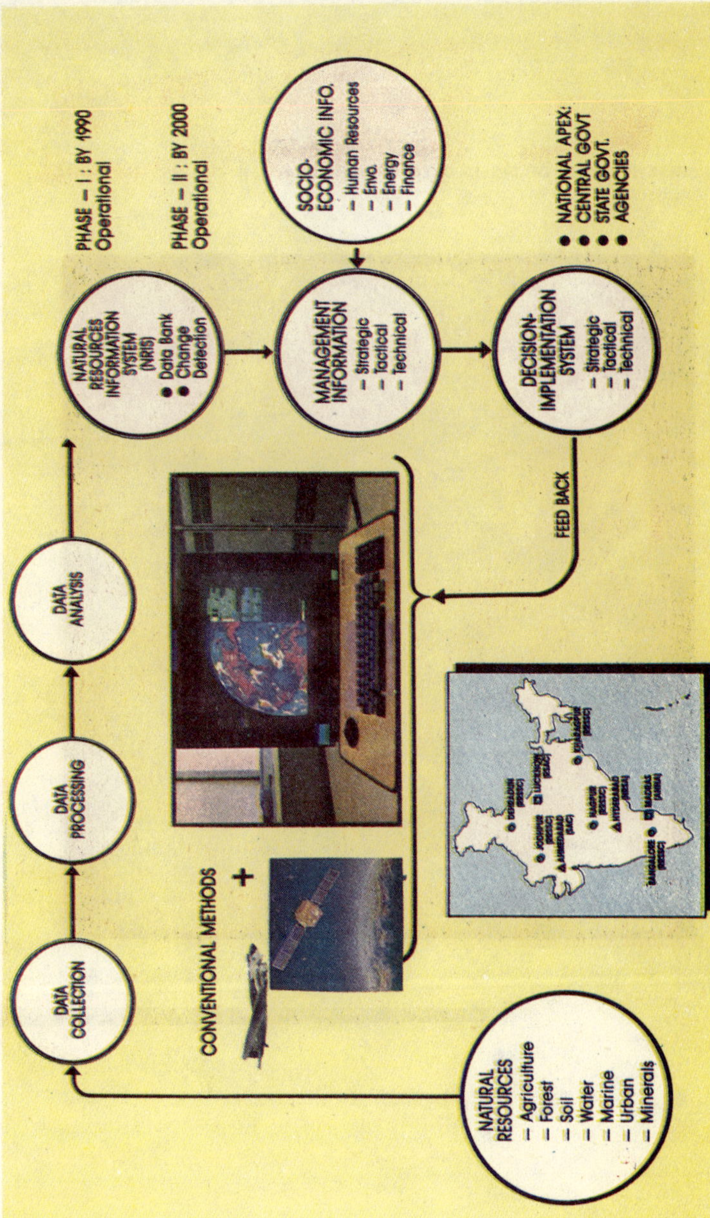

FIG. 7.1: National Natural Resources Management System

With Compliments From:
PUBLICATIONS & PUBLIC RELATIONS UNIT
INDIAN SPACE RESEARCH ORGANISATION
DEPARTMENT OF SPACE, BANGALORE.

The IRS-1 will be a series of 4 satellites 1A to D. IRS-1B similar to 1A will be launched during 1990-91. IRS-1C & 1D satellites will have enhanced capabilities by way of better resolution and repetivity.

Infrastructure Development

Towards proper utilisation of the potential of remote sensing, it is necessary that appropriate interpretation facility is available in states and union territories. At the instance of the Department of Space, several state and union territories of India have established remote sensing centres. Remote sensing centres in Tamil Nadu, Orissa, Uttar Pradesh, Rajasthan, Kerala, Madhya Pradesh, Bihar, Andhra Pradesh, Maharashtra, Gujarat, Haryana and Punjab are operational. The state governments of Karnataka, Manipur, Meghalaya, Jammu & Kashmir are in the process of setting up state remote sensing centres. All other states and union territories are planning to set up their own centres. The remote sensing centres in Tamil Nadu and Uttar Pradesh have procured large image processing systems built around VAX 11/780 computer. Haryana, Orissa and Madhya Pradesh state remote sensing centres and planning to procure medium level image processing systems. Operational state remote sensing centres have taken up a number of application projects. They are also participating in the national level projects such as nationwide wasteland mapping, ground water potential mapping etc. In addition to the facilities available at state remote sensing centres, a large number of user departments have set up their own remote sensing cells. They include central and state user departments and universities. Most of them have only visual interpretation equipments.

The Government of India have set up five Regional Remote Sensing Service Centres (RRSSCs) to provide sophisticated digital data processing facilities to the scientists of the user departments involved in mapping, inventory and monitoring of natural resources. In the initial years the RRSSCs will also offer training courses to the users on digital processing of remote sensing data besides offering regular user services. In addition to the digital analysis facility, each RRSSC will be equipped with a photographic processing facility, ground truth data collection equipment and limited visual interpretation equipment. Details of the facilities available in the country are given in Fig. 7.2.

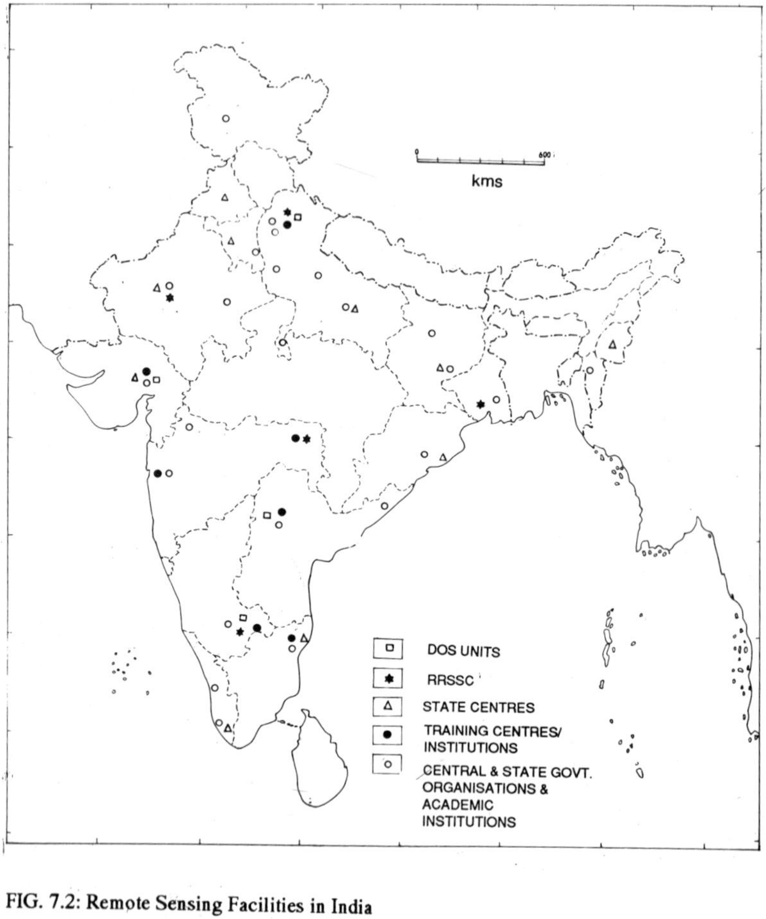

FIG. 7.2: Remote Sensing Facilities in India

Training

Training institutions such as the Indian Institute of Remote Sensing (Dehradun), National Remote Sensing Agency (Hyderabad), Space Applications Centre (Ahmedabad), Centre of Studies for Resources Engineering (IIT, Bombay), Institute of Remote Sensing (Anna University, Madras), Geological Survey of India (Hyderabad), Forest Survey of India (Dehradun), and National Bureau of Soil Survey and Land Use Planning (Nagpur) offer training courses in various natural resources themes of short and long durations for the benefit of remote sensing user community in India and developing countries. The Indian Institute of Remote Sensing, Dehradun is offering regular training courses at post-graduate level as well as short duration certificate courses. This Institute has so far trained over 1800 scientists and engineers in aerial photo-interpretation and remote sensing techniques. The courses offered by Centre of Studies for Resources Engineering and Institute of Remote Sensing include themes such as water resources, mineral exploration, terrain evaluation and land use planning.

Application Projects and Thematic Mapping

The remote sensing application studies were initiated in India by Department of Space through a series of end-to-end experiments in collaboration with several user departments and agencies. Under this programme as many as projects were selected covering various resource themes such as forestry, agriculture, water resources, soil, landuse, etc. Details of the projects and the salient results are given in Table 7.1. The results of the joint experiments were presented during the National Seminar held at Hyderabad during May 1983.

TABLE 7.1: Details of the End-to-end Experiments

I. **Agriculture and Forest Related Studies**

1. Crop acreages using satellite data (for major crops like wheat, rice and plantation crops, etc)
2. Crop yield forecasting
3. Soil mapping (including salinity of the fields)
4. Estimation of irrigated land and drought monitoring
5. Forest canopy area estimation including afforestation/deforestation estimates

II. Water Resources and Water Quality

6. Location of surface water and other water resources and their mapping; monitor their consumption
7. Watershed characteristics and hydrologic models
8. Regional flood mapping from space
9. Ground water exploration
10. Snow cover estimation
11. Littoral drift of east coast

III. Geological Surveys, Mineral and Petroleum Exploration

12. Detection and location of anomalies for potential sources of minerals and oil

IV. Cartography and Land Use

13. Updating of the existing maps at different levels including studies related to urban development

V. Environmental Monitoring and Protection

14. Water quality and colouration
15. Surface mining, jhuming, land/coastal erosion, deforestation
16. Salinity of large agricultural areas

VI. Marine Resources Oceanography and Coastal Engineering

17. Monitor shore-line changes

VII. Disaster Warning

18. Effective forecasting of floods

Note: The end-to-end experiments were carried out in active collaboration with the concerned user agencies and departments.

Indian Remote Sensing Satellite - Utilisation Programme (IRS-UP)

After the National Seminar it was decided to initiate a set of application projects called the Indian Remote Sensing Satellite - Utilisation Programme (IRS-UP) under which it was planned to provide the user community the experience and expertise to utilise the data that would become available from Indian Remote Sensing Satellite (IRS-1). The IRS-UP consisted of 16 projects selected in consultation with a number of users. The projects were subgrouped into four categories as follows :

Projects	*Subgroups*
Flood Mapping Regional Geological Mapping Ground Water Studies Land Use Mapping	Operational
Soil Mapping Land Degradation Snow Mapping Drought Monitoring	Quasi-operational
Crop Production Forecasting Forest Mapping & Damage Detection Water Quality Monitoring Watershed Characterisation	Experimental Application Project
Monitoring of Coastal Environment Marine Fisheries Experimental Application Projects Crop Stress Detection Crop Yield Modelling	Techniques Development Project

The operational and quasi-operational projects covered themes for which remote sensing application packages could be developed for routine, operational use by user agencies. The other two categories required specific R&D studies to bring them to operational status. In the IRS-UP projects there was participation by scientists from over 200 organisations at state and central levels. Almost all the states were represented. Under these projects a number of thematic maps have been prepared. Important among them are regional geological maps, ground water potential zones, landuse/cover maps, soil types, forest and coastal maps. Though the areas covered by these maps form only project sites covering parts of states, they proved the usability of remote sensing techniques in preparation of these maps (Fig. 7.3).

Remote Sensing Application Mission (RSAM)

Another major programme initiated by Department of Space, the Remote Sensing Application Mission (RSAM), has the main objective to prove the replicability of remote sensing techniques for different geographical and environmental conditions so that remote sensing could become a tool in thematic mapping at various levels such as regional, state and national. Under the RSAM, operational themes were selected to cover large parts of the country for mapping. The major themes covered are :

- Flood
- Urban Sprawl
- Crop acreage for major crops (rice, wheat and groundnut)
- Drought Monitoring
- Landuse/cover

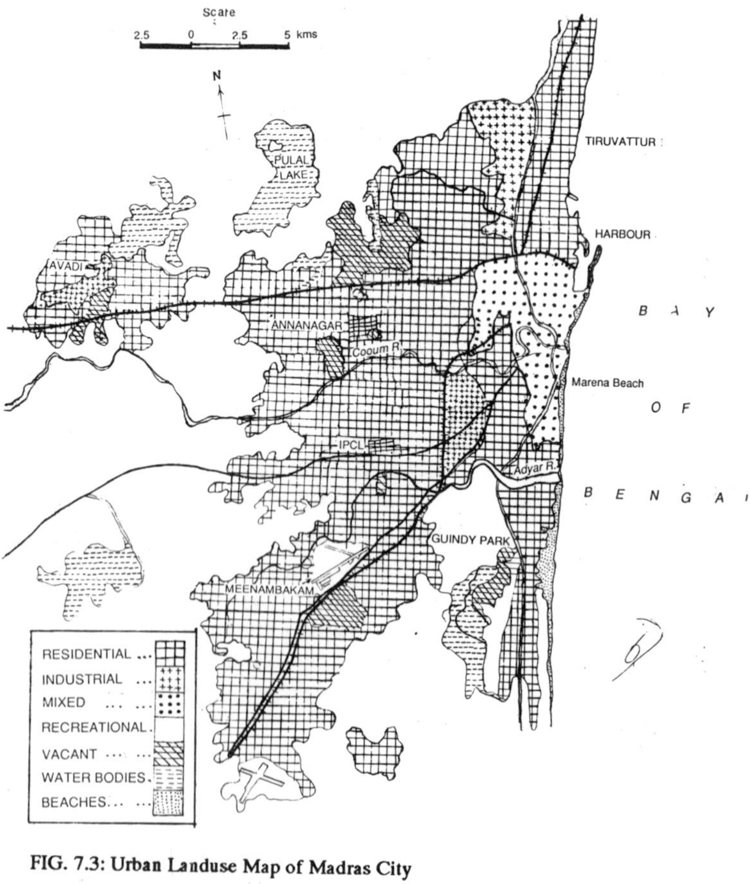

FIG. 7.3: Urban Landuse Map of Madras City

With the successful conduct of the RSAM, the applicability of remote sensing for diverse areas of coverage and a variety of application themes have been proven.

National level projects

In order to meet the requirements of the user departments and agencies several national level projects were carried out by ISRO/DOS. Forest mapping at national level was carried out for the two periods on 1:M scale in 1972-75 and 1980-82 by National Remote Sensing Agency. These maps gave three categories of forest cover, such as closed, open and mangroves based on the spatial distribution of forests. These maps provided, probably for the first time, the status of forest cover in the country and its trends in various regions. Presently Forest Survey of India has taken up the biennial mapping of forest cover for the entire country on a routine basis. The first cycle of mapping is nearing completion.

At the instance of the National Wasteland Development Board, mapping of 147 critically affected (selected) districts for wastelands were carried out. Thirteen categories of wastelands such as gullied/ravine lands, waterlogged areas, uplands with/without scrub, shifting cultivation, mining/industrial wasteland, degraded pastures/grazing land, degraded land under plantation crop, steep sloping area, land affected by salinity/alkalinity, sands (desertic/coastal), barren rocky/stony wastes degraded land under plantation crop, snow covered/glacial area were mapped on 1:50,000 scale using standard visual interpretation techniques. The work was carried out by 20 work centres spread over the country with NRSA as the nodal agency. Details of the districts covered under this project are given in Table 7.2..

Under the Drinking Water Technology Mission coordinated by Department of Rural Development, initially 56 mini-mission districts were selected for mapping of ground water potential zones. The geological and geomorphological features are identified from satellite data and the same is utilised to determine the potential for ground water availability. The mapping is carried out by the state and central user departments or agencies. The 1:250,000 scale maps are utilised for identifying potential zones for further geophysical investigations for targetting well sites. The inputs received from various states indicate that the success rate in drilling went up to 80-85 per cent as against the 45-50 per cent by only conventional means. It is planned to cover the entire 2.5 lakh problem villages under this mapping.

TABLE 7.2: Statewise Distribution of Districts Undertaken for Wasteland Mapping

State	Districts
Kerala	Kasargod, Cannonore, Palghat, Waynad, Idduki, Malapuram
Tamil Nadu	North Arcot, Periyar, Tiruchirappalli, Madurai, Dharmapuri, Salem, Ramanathapuram
Andhra Pradesh	Nizamabad, Vizianagaram, Karimnagar, Chittoor, Ranga Reddy, Medak, Cuddapah, Vishakhapatnam, Mahaboobnagar, Prakasam, Nalgonda, Kurnool, Anantapur, Nellore
Karnataka	Belgaum, Raichur, Bellary, Chitradurga, Hassan, Kolar, Tumkur
Maharashtra	Ratnagiri, Satara, Pune Sangli, Thana, Nasik, Kolhapur, Jalagaun, Ahmednagar, Dhule
Orissa	Gangam, Cuttack, Bhubaneswar, Dhenkanal, Sundargarh, Bolangir, Kalahandi
West Bengal	Midnapore, Purulia, Bankura
Bihar	Singbhum, Dhanbad, Nawada, Siwan, Hajipur (Vaishaly), Muzaffarpur, Rohtas (Sasaram), Palamau (Daltonganj), Aurangabad
Madhya Pradesh	Bhind, Raisen, Narasingapur, Indore, Shajapur, Ujjain, Mandsaur, Rajgarh, Rajnandgaon, Ratlam, Raipur, Durg, Duna, Datia, Vidisha, Dewar, Rewa
Gujarat	Gandhinagar, Bhavnagar, Kheda, Ahmedabad, Mehsana, Panghamahar, Surendranagar, Jamnagar, Junagadh
Rajasthan	Pali, Tonk, Jodhpur, Churu, Ajmer, Sawai-Madhopur, Bundi, Jhalawar, Kota, Bhilwara, Udaipur, Dungarpur
Urrar Pradesh	Fatehpur, Kanpur, Etah, Agra, Aligarh, Varanasi, Lucknow, Ghazipur, Azamgarh, Allahabad, Sultanpur, Jaunpur, Bela Pratapgarh, Hamirpur, Unnao, Mainpuri, Etawah, Jhansi, Lalitpur, Rae Bareli, Farukhabad, Jalaun
Punjab	Sangrur, Kapurthala, Bhatinda
Haryana	Sirsa, Karnal, Hissar, Jind, Gurgaon
Assam	North Cachar (Haflong), Karbi Anglog (Diphu)
Himachal Pradesh	Hamirpur, Dharmasala (Kangra), Chamba
Manipur	Senapati, Imphal
Nagaland	Kohima, Mokokchung, Zunheboto
Jammu & Kashmir	Udhampur

TABLE 7.4: Status of Thematic Mapping Under NNRMS

Theme	Thematic Output	Coverage	Remarks
1. Forestry	Forestry maps on 1: 250,000 scale	National & States	Updated every 2 years by FSI, DOEn&F
2. Ground Water	Ground Water Potential Zones on 1:250,000 scale	— Entire states of Maharashtra, Karnataka, Bihar Gujarat, Rajasthan & MP	Entire country to be covered under the Drinking Water Technology Mission coordinated by Dept. of Rural Development
3. Wastelands	Wasteland maps (13 categories) on 1:50,000 scale	146 selected critically affected districts	Work completed in March 1988 by NRSA in active collaboration with 20 work centres at state/central level
4. Soil	Soil maps on 1:1M scale state 1:250,000 scale	State and national level	Work being carried out by NBS & LUP
5. Urban Sprawl	Urban land use/cover maps on 1:50,000 scale	12 major cities (above 1 million population)	ISRO/DOS in collaboration with city development agencies and TCPO's
6. Landuse/cover	Landuse maps (showing agril/ forestry/soils) on 1:250,000 scale	Entire country covering 15 agroclimatic zones	Work being carried out by ISRO/DOS at the instance of Planning Commission with active involvement of states/central agencies

(Contd.)

TABLE 7.4 (*Contd.*)

Theme	Thematic Output	Coverage	Remarks
7. Coastal	Coastline maps on 1:50,000 — 1:250,000 scale	Entire Coast	SAC, ISRO in collaboration with coastal States
8. Floods	Flooded area maps on 1:250,000 scale	All major floods from 1986	NRSA has developed and transferred the methodology to user agencies
9. Derived Parameters			
Soil salinity/ alkalinity	Salinity/alkalinity maps on 1:250,000	Selected, affected districts in UP	Methodology developed by NRSA with user involvement
Vegetation Index	Vegetation Index (ratio of visible and near infrared) at 1 km resolution	Entire country at district level from 1986 Kharif season	VI maps being generated for drought monitoring
Flood prone/flood risk zones	Flood prone area maps on 1:250,000 scale	Selected areas in Uttar Pradesh	—
Water Quality	Water quality levels	Major lakes - Chilka, Pulicat and Dal	Methodology developed by SAC

Future Thrust Areas

With the experience gained over the past decade and half, several area of applications of remote sensing has become operational and semi-operational. The status of remote sensing applications to various natural resource themes is given in Table 7.5. A number of areas have been identified as thrust area for R&D so as to bring them to operational level in coming years. It is also planned to provide satellite data with better resolution, covering more spectral regions and higher repetivity so as to meet the user requirements towards generating thematic maps with better scales and information content. In order to overcome the difficulty of obtaining remote sensing data over persistant cloud covered areas, the microwave remote sensing programme has been initiated. The programme will consist of developing indigeneous sensors such as SLAR and SAR for use in future satellites. The projected plan of operationalisation of remote sensing applications in the near future is depicted in Fig. 7.4.

Conclusions

National Natural Resources Management System is conceived as an optimal management tool to monitor and manage our rich natural resources. The system will effectively integrate data or information available from conventional and remote sensing based methods. Under this NNRMS, being established by Department of Space, a large amount of thematic mapping work has been carried out covering various natural resources themes. Towards ensuring regular mapping of the natural resources, several major steps have been taken, such as development and launching of remote sensing satellites, reception and dissemination of data, infrastructure and facility built-up for data analysis and training of adequate manpower. It is hoped that with the establishment of NNRMS as an operational system in the country, the thematic maps covering various natural resources will become available routinely for use as an input to aid the development and planning process in the country.

Acknowledgements

The authors are grateful to Ms S. Vanaja for secretarial assistance and Ms M. Sameena for cartography work.

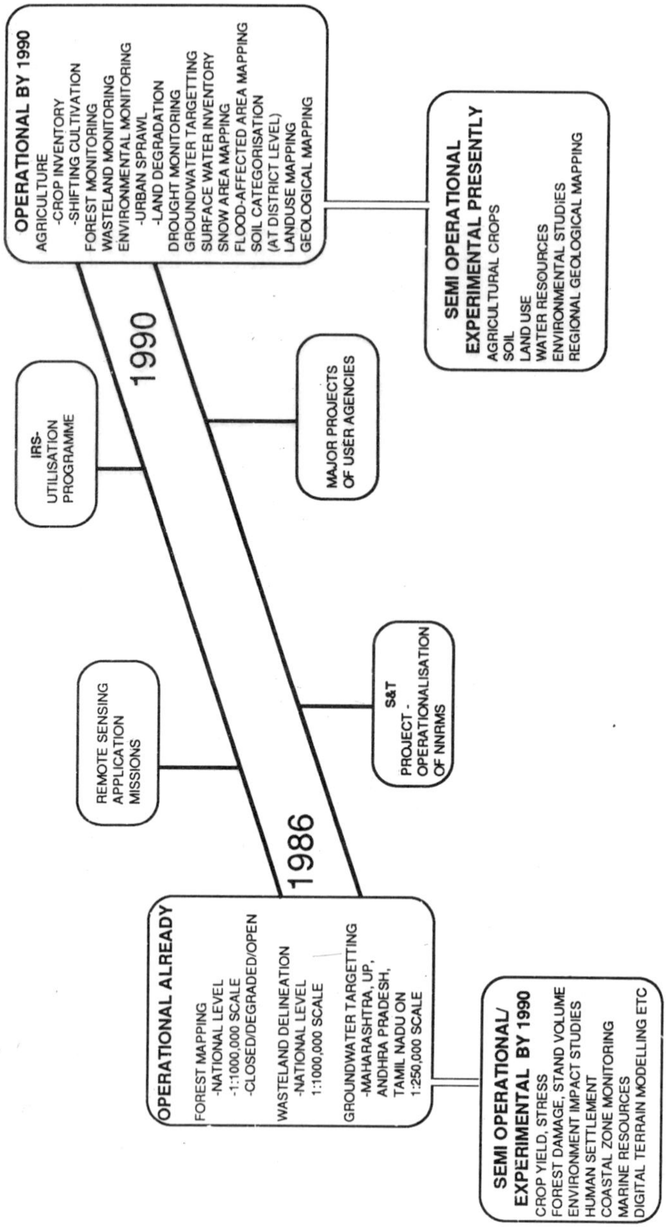

FIG. 7.4: Operational Remote Sensing Applications

TABLE 7.5: Status of Remote Sensing Applications

Theme	Operational	Semi-operational	Future Thrust Areas
Agriculture	* Crop Acreage	* Command Area Irrigation Management * Drought Monitoring	* Soil Mapping * Pest/Disease Damage * Crop Yield * Crop Stress
Geology	* Regional Geology	* Lithology, Rock Types * Mineralisation Zones	* Oil & Natural Gas Targetting * Mineral Targetting
Water Resources	* Ground Water Potential Zones * Surface Water Body * Flood/Snow cover Maps	* Watershed Characterisation * Reservoir Silting * Flood Damage	* Water Quality * Water Balance
Forestry	* Density & Type * Jhumming/Degraded Forests	* Stock Maps * Biosphere Reserves	* Volumetric Estimation * Forest Pest/Diseases
Ocean/Marine	* Nearshore Bathymetry * Sedimentation * Coastal Vegetation	* Sea State: Winds/Waves * Sea Surface Temp/ * Ocean Colour patterns	* Fishery Potential * Ocean Pollution

REFERENCES

Chandrasekhar, M.G., *et al.* (1988) Remote sensing activities in India, National paper presented at 9th ACRS, Bangkok, Thailand, Nov.

Bhavsar, P.D. (1980), Remote Sensing Activities in India, *Proc., First Asian Conference on Remote Sensing*, Bangkok, Thailand, Nov.

Proceedings of National Seminar on NNRMS, Vol. I & II, Hyderabad, May 1983.

Rajan, Y.S. (1984), A decade of remote sensing in India, some salient results presented at COSPAR, Graz., Australia, June-July.

8

ROLE OF REMOTE SENSING INSTRUMENTS IN EXTRACTING THEMATIC INFORMATION

J. KRISHNAMURTHY, A.S. PADMAVATHY AND M.G. CHANDRASEKHAR

Remote sensing technique for earth and environmental studies is one of the major application of space technology which has become an efficient tool for systematic survey, mapping, monitoring and better management of both renewable and non-renewable natural resources. The aerial remote sensing using single band cameras as well as multiband cameras are being extensively used for aerial mapping since few decades. Later, the advent of space based remote sensing techniques added new dimension to space technology which offers synoptic coverage of the area of study, timely acquisition of information, frequent repetitive/temporal observation of large and even inaccessible areas for making efficient developmental plan of natural resources by resources managers, planners and decision-makers. Several application studies covering various themes such as geology, landuse, water resources, agriculture, forestry, environment etc. are being carried out by user community using remote sensing techniques in conjunction with the conventional resources survey and mapping methods.

The thematic details can be derived from the remotely sensed data either through visual interpretation or digital data analysis. Both these human interpretation and machine—assisted processing of digital data demand suitable equipment ranging from simple magnifier to sophisticated interactive computer system. Selecting the appropriate system for data handling depends upon the user specific requirements.

Remote Sensing Technique

Remote sensing is the technique used to obtain information about the earth surface and atmosphere without having physical contact with the object of study, usually by means of radiation from the electromagnetic spectrum. Depending on its physical features and chemical properties, the earth surface emits different amounts of electromagnetic energy useful for various interpretation and analysis for extracting details of interest which forms the basis of remote sensing technique. The data can be collected by the remote sensing devices including passive and active systems and employ different bands in the visible, near infrared, middle infrared and far infrared as well as in centimetre radiowaves. In passive remote sensing, the reflected or emitted electromagnetic energy is measured by sensors operating in different selected spectral bands where the original source of energy is the Sun. But in the case of active remote sensing method the earth feature is illuminated by an artificial source of energy. Thus the emitted and reflected energy/signal detected by the sensors on board platforms are transmitted to the earth station. The received signal is then processed in a variety of ways based on the user's requirements.

Satellites and Sensors

Space scientists realised the potentials of orbiting space platforms since the launch of the very first satellite, the Russian "Sputnik" in 1957. During sixties a series of satellites like "Apollo", "Gemini" etc. were made available to resources scientists. However, remote sensing application especially for earth resources and environmental studies came into its own from 1972 with the launch of *Earth Resources Technology Satellite*, later renamed as Landsat by National Aeronautics and Space Administration (NASA). Since then a number of satellites have been placed in orbit by various countries like USA, Canada, France, Japan, India. Presently the Landsat 5, the fifth satellite in the Landsat series by NASA, SPOT satellite by French Space Administration, indegeneously developed Indian Remote Sensing Satellite — 1A (IRS-1A) by India are orbitting the earth with specified period of interval and continuously provide multi-temporal and multi-band data. The sensor payloads of these satellites vary in respect of spectral bands, spatial resolution and radiometric resolution. While the SPOT satellite is unique in providing stereoscopic images, the sensors of IRS-1A provides the images of the same region in same spectral band but in two different spatial resolutions of 72.5 m and 36.25 m.

Types of Data Products

The satellite data products are available to the resources scientists for extracting required thematic information in the form of (a) photographic products, and (b) digital form as computer compatible tapes (CCTs). The photographic products such as paper prints, film negatives, diapositives of black and white and false colour composite are available from 1:1 million to 1:25,000 scale. The digital tapes are having density of 6250 bpi or 1600 bpi. Data products are also available either as raw data or preprocessed data after carrying out necessary corrections like sensor radiometry, earth rotation, earth curvature, platform altitude and orbit related parameters.

Image Interpretation and Analysis

Image interpretation and analysis is the process of detection, recognition and classification of earth's features. Extraction of thematic details from remotely sensed data can be either through visual interpretation or digital data analysis. Obtaining details of land cover or morphological features from satellite based data are relatively simpler than other applications such as geological, hydrological, forestry and ecological, pedological where details have to be inferred directly or indirectly through the surface manifestations. The photo-interpretation keys such as tone, texture, pattern, association etc. are being used by the interpreter while analysing the image visually. Subsequent to the introduction of digital image processing techniques for thematic mapping from remotely sensed data, newer methods of data handling, processing and classification have been designed. Image enhancement and classification are the two major components of image processing techniques. Even though the human interpretation technique cannot be replaced totally, at least a partial replacement is justifiable due to (a) for coping with increased and voluminous data flow, and (b) flexibility in data manipulation. The improvements in data analysis can lead to both faster and more accurate extraction of thematic details. Thus the required thematic output depends on the type of methods used for image interpretation and analysis whether it is human interpretation techniques or machine-assisted processing methods.

Remote Sensing Instruments

The visual interpretation and digital analysis of remotely sensed data for extracting thematic details of interest requires suitable equipment ranging from simple magnifier to interactive digital image processing system. In addition, there is a need to have support facilities to meet the cartographic and ground truth data collection.

Based on the type of remotely sensed data used as input, the remote sensing equipment can be broadly grouped into two categories, viz. (a) visual interpretation and information transfer aids when photographic products are used as inputs, and (b) digital image processing system when the data stored in computer compatable tapes are used for analysis.

Visual Interpretation and Information Transfer Aids

Based on the purpose of use, the visual interpretation and information transfer instruments can be broadly grouped into three categories viz.

(a) *Instruments useful for viewing aerial photographs and satellite images either stereoscopically or non-stereoscopically*: Light table, viewing desk are examples of simple aids useful for tracing the required thematic details from photographic products. Stereoscopes of various types are used to view the terrain three dimensionally. Images can also be viewed at larger scale during visual interpretation using magnifiers, large format optical enlargers etc. Optical colour combiner like additive colour viewer and electronic colour combiner like image analyser are useful to view the image with certain specific enhancement.

(b) *Measuring instruments*: Instruments useful for linear, area, height and densiometric measurements.

(c) *Transfer equipment*: instruments that record or transfer details based on either *camera lucida* principle or employing optical projection. These instruments are used essentially for transferring the details to the required scale. Instruments such as transferoscope, stereozoom transferoscope, vertical sketch-master uses *camera lucida* principle in which the eye receives two superimposed images, one from the photograph and another from a base map. Optical reflecting projector, map-o-graph, pantograph etc. employ principle of optical projection, where the images of photographic prints project through a lens and mirrors onto a mapping surface.

The broad description of some of the visual interpretation and information transfer instruments are given below; there are various versions/improvements of instruments by different manufacturers which may differ in sophistication and flexibility of operations.

Light Table

Light table is a basic instrument required by any image/photo interpreter for studying remote sensing photographic products of paper prints and films. The uniform illumination throughout the entire observation area helps the user to differentiate the various features in same illumination condition. The variable intensity helps to study the different kinds of data products since the density of the photographic material may vary.

Illuminating Magnifier/Magnascope

It is a simple optical device useful to view the input photographs/ images during the visual interpretation on an enlarged scale.

Pocket or Lens Stereoscope

It is a type of stereoscope consists of two simple convex lenses mounted on a frame useful for viewing stereo pairs. It has the provision for changing the spacing between the lenses to accommodate various eye bases. The legs can be folded or removed so that the instrument is easily stored or carried to field.

Mirror Stereoscope

Mirror Stereoscope is a type of stereoscope consists of mirrors in addition to simple lenses in order to suitably view the large area of each pair of photographs or images. Since this instrument permits the two photographs on images to be completely separated while stereoscopic viewing eliminates the problem of one photo/image obscuring part of the other. This instrument has the provisions for changing eye bases or binoculars.

Scanning Stereoscope

Scanning Stereoscope is a type of an instrument which has suitable provision to scan the stereomodel in both X and Y directions without moving the photographs/images or orientation of the stereoscope. The photographs/images are viewed through oculars and they can be individually focussed and magnified.

Zoom Stereoscope

Zoom Stereoscope is essentially developed for stereoscopic analysis of photographs/images of negative or positive film or plates. This instrument has provision for continuous change in magnification accomplished by the zoom lens system and focussing and rotation of field of view. These facilities allow the operator to view the image at the best possible magnification for observing the maximum image details.

Additive Colour Viewer

Additive Colour Viewer is a simple optical colour-combining equipment for generating and viewing colour composites by introducing required colour combination from black and white multispectral transparencies of aerial or satellite images. The multispectral photographs and images taken simultaneously can be used for generating true colour composites or false colour composites by using coloured filters blue, red and green, or their complementaries yellow, cyan and magenta. By varying colour, density and hue the interpreter can easily distinguish subtle details (Plate 8.1).

Image Analyser

It is an electronic colour-combiner specially designed to provide colour composites as well as density sliced image display from black and white transparency or hard copy for interpretation and analysis. This system can be effectively used to get the desired thematic information from black and white images by assigning the colour tones of varied range and intensity to each gray level. The colours assigned to the density zones are red, yellow, orange, green, violet, cyan, blue and magenta. It is also possible to calculate the area of each colour density zones for classification of different features (Plate 8.2).

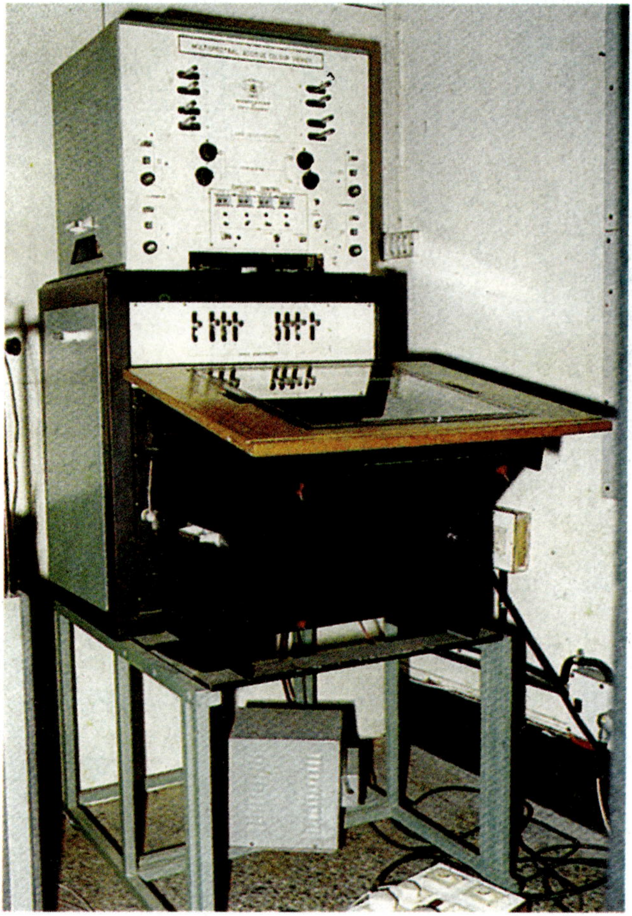

PLATE 8.1 : Additive Colour Viewer

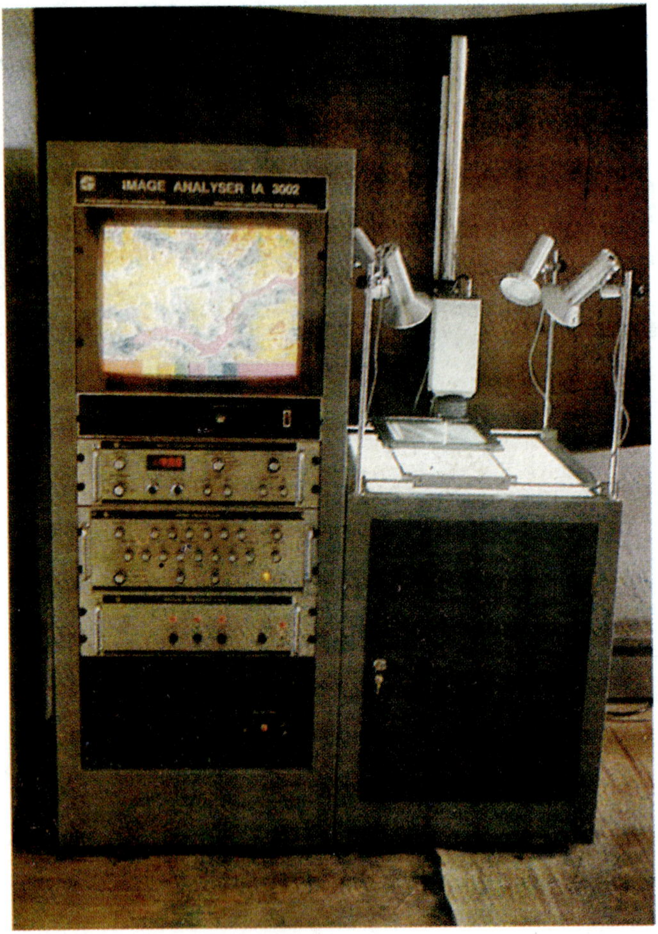

PLATE 8.2 : Image Analiser

Dual Densitometer

Dual densitometer is basically a photometer provided with digital readout designed to measure the transmission and reflection densities of photographic films and prints of both black and white and colour products.

Sketch Master

Sketch master is a simple instrument which provides a superimposed image of an aerial photograph/satellite image on to a map. This instrument has the provision to raise or lower the viewing unit on its support to adjust the photo scale to that of the map. Also the frame on which the photo or image is mounted can be tilted on a swivel joint so that rectification of the parts is possible to correspond to the map.

Stereosketch

Stereosketch is a type of sketch master that provides the superimposed views of a base map and a stereoscopic model of a stereo pair. The stereoscopic feature allows an appreciation of relief which is useful in interpretation. Also the planimetric errors due to parallax difference are reduced. This instrument can be used for plotting formlines and for hillshading (where the direction of illumination is assumed to be form the northwest slopes are shaded accordingly). This instrument can correct tilts up to 5° and provides means for varying the plotting scale.

Zoom Transferscope

Zoom transferscope is useful for transferring the graphic information from a photograph or image to the base map when both are at different scales. When the photo scale is different from the map scale, the zoom control enables to magnify or reduce the photograph until it matches the scale of map. Its anamorphic optical system enables different magnification ratios to be applied in the X and Y directions. The zoom magnification provides continuous variations in magnification from 1 to 7X. It also has an image rotation system which operates by means of rotating prisms. The photo input to this system can be either transparent or opaque photographs in black and white or colour. Satellite images can also be used with certain maps and charts.

Stereo Zoom Transferscope

It is a versatile optical instrument used for planimetric map revision. Using this, the graphic information from satellite and aerial photographs can be transferred on to the base map when both are at different scales. This instrument can be operated in three different modes, viz. stereo, mono or photographic mode. The stereoscopic provision offers three dimensional viewing of the photo pair, superimposed on the map. Its anamorphic optical system enables different magnification ratios to be applied in the X and Y directions. It has zoom magnification providing continuous variations in magnification from 1 to 7X separate zoom optics (range 7:1) for each image enables simultaneous study in combination, of images of diverse scale, date and type. In addition, it has an image rotation system which facilitate adjusting photo image to coincide with map points on control points. The photo input to this system can be either transparency or hard copies of black and white or colour photographs of 9´´×9´´ format. Satellite image can also be used with certain maps and charts.

Plan Master

Plan master is an optically variable copying instrument essentially designed for transfer of information from transparent or opaque original on to map manuscript at required scale.

Optical Reflecting Projector

Optical reflecting projector is a simple optical projection system useful to enlarge or reduce the input photographs, maps or drawings to the required scale. This system can generate distortion and vignette free images for desired scale factor. The input drawing or photograph is illuminated by the illumination units and projected on to the plane. The projected image can be traced on a transluscent or semi-transparent sheet. The major advantage of the system is that it simplifies manuscript preparation, compilation, base map work, interpretation, revision or updating any detail involving scale variations (Plate 8.3).

Optical Pantograph

Optical pantograph is a latest rapid scale change drafting and interpretation instrument. This instrument is designed specially to obtain geometrical rectification or distortion corrections while transferring the data from aerial photographs for satellite images to maps by providing motorised tilt in X and Y directions to copy platform. This system can accept large farmat inputs, continuously variable enlargement and reduction from 4X to 1/4X.

Large Format Optical Enlargers

Large farmat optical enlarger is very similar to a photo enlarger and utilises the principles of optical projection to form enlarged images at variable magnification for easy visual interpretation of black and white or colour composite transparency images (Plate 8.4).

Digital Image Processing System

Simple interactive mini image processing systems to *main frame* interactive image processing systems are used for processing and analysing of remotely sensed digital image data (Plate 8.5). A variety of programmes are available to perform the digital image processing techniques to obtain the desired thematic details. There are three major components in digital image processing of remotely sensed data viz. (a) pre-processing (b) image enhancement, and (c) classification. It is very difficult to have a cut off line between preprocessing and image enhancement since both will supplement and complement each other. In general, corrections carried out for atmospheric effects, earth rotation, altitude and attitude forms part of preprocessing of image processing techniques. The image enhancement is being carried out by a set of transformation and manipulation functions to facilitate analysis and classification. However, classification is a process by which the image being assigned to particular categories or classes.

Pre-processing

Pre-processing is a technique adopted for creation of more accurate depiction of the image in terms of its radiation characteristics and geometric properties. The raw multispectral scanner data may have two types of errors which include random and systematic sources of errors. The random errors are difficult to be detected because of the very nature of the error. Whereas, the systematic errors can be detected

PLATE 8.3 : Optical Reflecting Projector

PLATE 8.4 : Làrge Format Optical Enlarger

PLATE 8.5 : Image Processing System

and corrected. The errors which require preprocessing include errors caused due to (a) platform effects, (b) sensor effects, (c) scene effects, and (d) atmospheric effects. In pre-processing, the radiometric correction is applied to obtain accurate radiance values for each picture element in the scene and geometric correction is applied to obtain accurate geometric representation of the earth.

Image Enhancement

Image enhancement in digital image processing technique is performed to improve the contrast of an image as well as to manipulate the data in a variety of ways by which to get the details of interest. A variety of programmes are available for image enhancement include stretching (contrast, gaussian, exponential, log, piecewise etc.), addition, subtraction, ratioing, edge enhancement, filtering, principle component analysis etc. Image contrast can be greatly improved by performing stretching function depending wherein spectral distribution are considered. The bands are added if different types of information is to be seen on the same image. Subtraction may be used to elimate the common features and bring out the difference information. The ratioing is done to get better contrast between or within classes by dividing the spectral value of a pixel in one band by that in the other. Principle component analysis is a powerful redenency reduction technique useful to compress the variance of the original data into the lower-order components.

Classification

Classification is a process by which the image being assigned to particular categories or classes. There are two types of classification, viz. (a) supervised classification, and (b) unsupervised classification. In supervised classification, classification is performed by a prior knowledge of the area by assigning training sets obtained from ground truth information, on the other hand the unsupervised classification is performed when there is no prior knowledge about the area is known. The unsupervised classification approach involves training the computer using *clustering technique.*

Development of Remote Sensing Equipment by Indian Space Research Organisation/National Remote Sensing Agency

Under technology transfer scheme of Department of Space the know-how for a variety of visual interpretation instruments, digital analysis systems, ground truth and cartographic instruments are being developed by units of DOS like NRSA, Hyderabad; Space Applications Centre (SAC), Ahmedabad; and ISRO Satellite Centre (ISAC), Bangalore and the same is transferred to Indian manufacturers for regular commercialisation. These instruments are essentially optomechanical, optoelectronic and microprocessor based image analysis systems and they are useful for various purposes like interpretation, analysis, transfer of details in varied scales, measurements of photo/image coordinates (ground control points), generation of different type of raw or processed data products, enhancement of images, checking the quality of data products, studying the changed detection, collection of ground truth data products, studying the changed detection, and collection of ground truth data.

The inputs for these systems vary from single band data to multiband data of various formats-diverse photographic products and digital image data stored in magnetic tapes. The outputs possible through these instruments range from format displays on the screens or display panels, tracings, drawings, colour coded plots, various photographic products and also computer compatable tapes. Qualitative and quantitative measurements can be made using these instruments.

The expertise required for manufacturing of these low-cost equipment has been transferred to suitable Indian industries for regular production, marketing and servicing to meet the growing users' demands. The know-how of remote sensing utilisation equipment has already been transferred to industries and they are in regular production (Table 8.1). The equipments which are in the pipeline and are either awaiting completion of development or licencing is given in Table 8.2.

National Natural Resources Management System

Realising the potentials of remote sensing in resources management, the Government of India approved the establishment of National

Natural Resources Management System (NNRMS). The NNRMS is a hybrid information system consisting of an optimum mix of remote sensing based systems as well as conventional systems. Department of Space has been identified as the nodal agency for speedy implementations of NNRMS in the country.

TABLE 8.1: Low Cost Remote Sensing Utilisation, Ground Truth Collection Cartographic Instruments/Systems and Licencee/Industries

Sl. No.	Instrument/Process/System Price (Ex-factory)	ISRO/NRSA Licencee Industries
1.	Optical Reflecting Projector	Optomech Engrs. Pvt. Ltd. Shed No. 3A, Type II Kukatpally, Hyderabad 500037.
2.	Diazo Printer	- do -
3.	Large Format Optical Enlarger	- do -
4.	Multiband Ground Truth Radiometer	- do -
5.	Map Master	- do -
6.	Automatic Microfiche Camera	- do -
7.	Multispectral Colour Additive Viewer	Remote Sensing Instruments H.No. 2-2-18/18/3/16, Bagh Amberpet, Hyderabad 500013
8.	Light Table with Co-ordinate Measuring System	- do -
9.	Simple Light Table	- do -
10.	Viewing Desk	- do -
11.	Colour Composite Printer	- do -
12.	Illuminating Magnifier	Dynascan Inspection Systems 19-10A, Cambridge Road Bangalore 560008.
13.	Analog Image Analyser	Speck Systems B-49, Electronic Complex Hyderabad 500762.
14.	Dual Densitometer	- do -
15.	Photowrite System	- do -
16.	Drum Pantograph	- do -
17.	Optical Pantograph	Hi-Tech Optics H.No. 2-2-18/3/1 Bagh Amberpet Hyderabad 500013
18.	Spectroradiometer	Hind High Vacuum Site No. 17, Peenya Indl. Estate, Bangalore 560058.
19.	Agrophotometer MK-II (Photometer for in-situ measurement of water stress and chlorophyll content in plant foliage)	Optomech Engrs. Pvt. Ltd., Shed No. 3A, Type II Kukatpally, Hyderabad 500037 ELICO, Sanatnagar Indl. Estate, Hyderabad 500018
20.	Hydrophotometer	Tata Tea Limited PO Box No. 9, Munnar 685612, Kerala

(Contd.)

TABLE 8.1 (*Contd.*)

Sl. No.	Instrument/Process/System Price (Ex-factory)	ISRO/NRSA Licencee Industries
21.	Interactive Intelligent Image Graphic Display Terminal	Sita Electronics, 12-5-35/A/9, Sriniket Tarnaka, Hyderabad 500017.
22.	Microprocessor based Multispectral Interactive Data Analysis System (Micro-MIDAS)	- do -
23.	Microfiche Camera	Micro Documentation Insts., 2-2-18/27, Besides Water Tank, Durgabai Deshmukh Colony, Bagh Amberpet, Hyderabad 500013. Optomech Engrs. Pvt. Ltd., Shed No. 3A, Type II Kukatpally, Hyderabad 500037.
24. to 28.	Optocoat-1,2,3,5,6	Optical Ccating Laboratory 16, 3rd Main Road Jayamahal Extension Bangalore 560048
29. to 31.	ISROVISION with SACIMAGE VIBGYOR-2001 (PC-AT) 386/ 387 based image analysis system) (ISROVISION with IBM PC/AT 386/387 co-processor, 8 MB memory Magnetic tape unit, printer, 2x40 MByte disk, video display processor and monitor with 4x1Kx1Kx8bit refresh memory planes etc. 35mm hard copy unit, digitizer, plotter etc., will be optional.)	HCL Limited, 806-808, Siddhartha 96, Nehru Place, New Delhi 110019. Speck Systems Pvt. Ltd. B-40, Electronic Complex Hyderabad 500762.
32.	High Magnification Enlarger	Machine Products India Private Limited Plot No. 41, GIDC Estate, Odhav Road Ahmedabad 382415.

The NNRMS Secretariat at ISRO, Bangalore is functioning with the main objectives of liaison with different users of state or central government departments or organisations and internal agencies in different tasks related to use of remote sensing data such as establishment of infrastructure for remote sensing centre, unit or cell, supply of equipment for data processing and analysis, and organising the information systems for use of remote sensing data in management of natural resources in the country.

TABLE 8.2: List of Low Cost Remote Sensing Data Utilisation Instruments/System/ Ground Truth Equipment Under Pipe Line/Awaiting Completion of Development and/or Licencing to Suitable Industries.

Instrument/System	Application/Utility
1. NRSA Transferoscope	transferring graphic information from a photograph to the base map when both are on different scales.
2. NRSA Stereo Transferoscope	transferring graphic information from a stereopair of photographs to the base map when both are on different scales.
3. Light Pointer	an aid to point out a location on a projected image on the screen without obstructing the image.
4. Synchronised Dual Visual Projection System and Automatic Audio Synchronisation	protects slides and tape audio inputs to present in synchronised fashion for lectures, seminars, presentations etc.
5. Texture Meter	a visual interpretation aid to study different types of surfaces by classifying them according to their textural characteristics.
6. 70 mm Film Viewer	viewing and interpreting 70 mm Landsat chips/role films at 1:25,000 scale
7. Aerial Film Drier	drying or processed wet aerial films in a heat controlled chamber without mechanical damages; scratches or distortion to the film on settling dust over the film.
8. Quick Look Film	10 x enlargement of aerial viewer films
9. Modified B2	2 x to 5 x enlargement of 70 mm format aerial films.
10. Variable Magnification Large Format Optical Enlarger	enlarge the black and white or colour composite transparency images at variable magnification for visual interpretation.
11. Portable Helium Analyser for Geological Surveys and Mineral Prospecting	measure the helium concentrations in any gas and liquid mixture in situ/field analysis.
12. Infrared Thermometer	measure the source temperature with reference to the ambient temperature and also to measure emmissivity.
13. Variable Angle of incidence Spectrophotometer	measure the optical properties of vegetation in live state, and also to derive the photosynthetic activity micro-nutrient effects on the growth of plants and crop condition.
14. Computer software for Contour Mapping	FORTRAN package contain subroutines useful for drawing contours under various complex situations-randomly/arbitrarily spaced data.

Conclusion

Various application studies carried out by different user departments through a small scale pilot studies to major national missions have led to the abundant use of various types of instruments some of which are discussed in this chapter. It is very difficult to imagine the scenario if these instruments or systems were not made available to the analysts. It is gratifying to note that most of the instrument are available indigeneously manufactured by Indian entrepreneurs, the quality of which are comparable and sometimes better than those available in elsewhere in the world. Presently efforts are pressed in the development of an expert system and also introduction of artificial intelligence in image processing of remotely sensed data. The Indian technologists, scientists and also industries are committed to the remote sensing community in making available state-of-the-art systems and instruments for the better utilisation of the remote sensing data.

Acknowledgements

The authors express their sincere gratitude to Shri K.R. Sridharamurthy for providing the details on low cost instruments developed by ISRO and NRSA. Acknowledgements are due to Ms S. Vanaja and Ms Suniya for secretarial assistance.

REFERENCES

ASP (1981), *Manual of Remote Sensing*, American Society of Photogrammetry, Falls Church, second edition, chapter 24, pp 987-1118.

Krishnamurthy, (1986) J. : *Remote Sensing Data Utilisation Facilities in India*, NNRMS-BR-3, September.

Krishnamurthy, J. (1987): *Role of Indian Space Research Organisation and National Remote Sensing Agency in the Development of Low Cost Remote Sensing Equipment in India*, ISRO-NNRMS-SP-32-87.

9

APPLICATIONS OF IRS DATA FOR THEMATIC MAPPING

S. JONNA AND S.K. SUBRAMANIAN

One of the primary objectives of the Indian space programme is to harness space technology for obtaining real time, accurate and reliable information on natural resources and resulted in operational satellite based remote sensing system, i.e., the Indian remote sensing satellite (IRS) which is generating direct or indirect information in various fields like geology, forestry, hydrology and agriculture. The first in the series of operational remote sensing satellite launched in 1988 was designated as IRS-1A with follow up satellites IRS-1B and IRS-1C. The next generation satellite IRS-2, may be augmented to include a number of new spectral channels covering middle infrared and thermal infrared portions of electromagnetic spectrum. Alongside, in another class of satellites, microwave pay loads (SAR with 25m spatial resolution) will be carried to enable the users to get integrated information at different spatial resolutions and in different spectral ranges to meet a variety of application needs.

Architecture of IRS-1 A

The IRS-1A system consists of a three axis body stabilised space craft (Figs. 9.1 & 9.2) designed for an operational life of 3 years in polar sun synchronous orbit at an altitude of 904 kms. The satellite spans 11 with both solar panels developed and weights 975 kg. The space craft is equipped with two deployable sun tracking, three panel rigid solar arrays to generate a power of 709 watts (EOL): TTC (Telemetry Tracking and Command) and data handling systems in S and X bands; altitude and orbit control system based on reaction wheels, magnetic torquers and monopropellant hydrozine based reaction system to achieve pointing accuracies of 0.4 degree and low drift rate of 3 × 10-4 degree/sec and jitter of 3 × 10-4 degree; the light weight aluminium honey comb panel based structure with active/passive thermal control elements. The detailed specifications of IRS-1A satellite given in Table 9.1.

TABLE 9.1: Overall specifications of IRS-1A Observatory

Type	: Body stabilised remote sensing satellite
Orbit	: Sun-synchronous, 904 km. orbit with equatorial Crossing at 10.25 a.m., descending node
Mission life	: Three years
Mechanical Configuration	
Structure	: Aluminium and aluminium honeycomb structure
Weight	: Approx. 975 kg.
Thermal Control	
Thermal Components	: Passive control using tapes, paint, OSR, MLI blankets and semi-active/active control using proportionate temperature controller and heaters
Temperature Range	: 20 ± 5 deg. C range for payload electro-optics 5 ± 5 deg. C for chemical batteries 0 to 40 for electronic packages
Payloads	
Type	: LISS (2048 elements CCD linear array)
No. of Cameras	: Three (One of LISS-I & two of LISS-II)
No. of Spectral Bands	: 0.45 — 0.52 microns 0.52 — 0.59 microns 0.62 — 0.68 microns 0.77 — 0.86 microns
IFOV (Micro-rad)	: 80.2 for LISS-I 40.1 for LISS-II
Spatial Resolution (m) (geometrical)	: 72.5 for LISS-I 36.25 for LISS-II
Swath	: 148.48 km. for LISS-I 146.98 km./for LISS-II (Combined swath)
Repetivity Cycle	: 307 orbits/22days
No. of grey levels	: 128
Attitude and Orbit Control	
Attitude Sensors	: IR horizon sensors (conical and static), star sensors, sun sensors, gyro
Attitude Control Thrusters	: Reaction wheels (4), magnetic torquers, hydrazine thrusters
Orbit Control Accurators	: Monopropellant hydrazine thrusters
Orbit Determination Accuracy	: 1 km.
Attitude Determination Accuracy	: 0.1 deg.

(Contd.)

TABLE 9.1 *(Contd.)*

Telemetry Tracking & Command (TTC)	
Telemetry	: HK information in S band; PCM/PSK/FM Real time rate 256 bps and play-back rate 4 Kbps. Onboard storage capacity of 104 min. of HK data
Telecommand	: S-band (PCM/PSK/FM/PM), and VHF (PCM/FSK/AM) ON/OFF commands 20
Tracking	: S-band tone ranging and two way doppler (major tone 100 KHz, minor tones 20 KHz, 4 KHz, 800 Hz, 160 Hz) X-band beacon and single way doppler
Power	
Solar Array	: 8.5 sq. metres area, deployable and sun-tracking generating 709 watts at EOL
Battery	: 2 Ni-Cd batteries of 40 AH capacity each

The sensor payload system consists of two pushbroom cameras (LISS-II) of 36.25 m resolution and one camera (LISS-I) of 72.5 m resolution employing linear charge coupled device (CCD) arrays as detectors. Each camera system images in four spectral bands in the visible and near IR region. The ground swath for the image obtained by the LISS-I camera is 148.48 km, whereas the LISS-II cameras image adjacent swaths of 74.24 km. width each, with an overlap of 1.5 km. across track. The reflected energy sensed for each picture element by the detector in quantised into 128 gray levels.

Nature or IRS-1A Data

There are four levels of data products available which are briefly dealt here.

1. *Quicklook and brouse products (Level 1)*

Quick look film and brouse films are generally used for monitoring the quality of data received. The brouse products are corrected for radiometric errors and are produced in about 48 hours of acquisition. Hence they can be used in emergency situations like flood area mapping.

2. *Standard products (Level 2)*

These products are generated after making necessary corrections for system errors like sensor radiometry, earth rotation, earth curvature, platform altitude and orbit related parametres. These products are also adjusted for map projection schemes adopted by Survey of India (SOI).

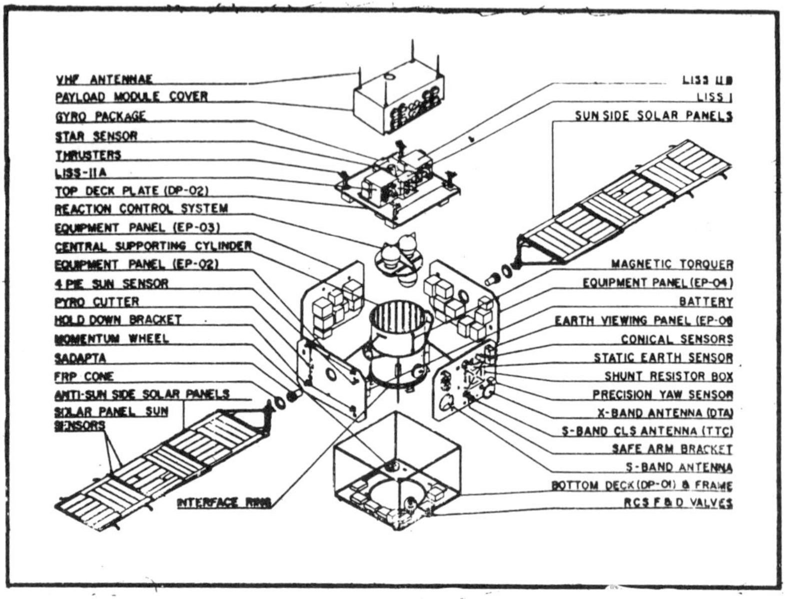

FIG. 9.1: Dis-assembled View of IRS-1A Satellite

3. Precision products (Level 3)

These products are generated after improving the location accuracy of imagery to better than 100 using ground control points.

4. Spatial products (Level 4)

(i) Enhanced products with contrast stretching, principal component transformations, ratios, etc. applied to a full scene or a subscene are available on request.

(ii) Geocoded products formulated with respect to a standard Survey of India topographical sheet at 1:50,000 and 1:250,000 scale based on the standard referencing scheme of SOI are also available at level 2 (2.2km rms error) or level 3 (100 m rms error)

All the above products are available to users in different scales in black and white colour photographic prints on transparencies form and on computer compatible tapes/floppies/cassettes. For users who are interested in carrying out corrections on the data on their own, it is possible to give the raw data and the brouse data on CCT. The turn around time for supply of data depends on demand for other data, priority and level of the product requested. The recent price list (subject to change) for various data products, data ordering information, terms for supply are given in Annexures.

Choosing the Products

Standard products way be selected based on the IRS referencing scheme (available with NRSA data centre, even on regional basis). It is advisable to specify with path and row number, the region of interest and also the latitude/longitude coordinates, since IRS path can create large lateral shifts in East-West direction. The required dates of imaging for the specific paths (scenes) can be planned from orbital calendar.

Geocoded Products can be ordered by map numbers and dates. Map sheets which fall on more than one scene will require multiple imaging since mosaicing is not yet offered. Standard geocoded products will have 5-6 pixel internal distortion which is not compatible to mapping standards of 1:50,000 scale, and so precision geocoded products are required for such purposes.

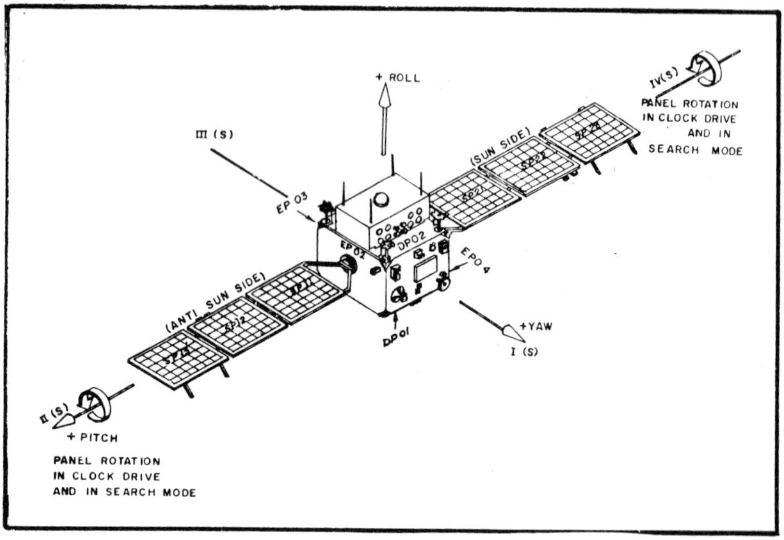

FIG. 9.2: IRS-1A Spacecraft

Special products can be ordered to cover areas as small as 512 × 512 pixels/scan lines (18 km × 18 km for LISS-II). Such products are ideally suited for intensive test sites and for urban studies. These products are useful for those who do not have access to digital image processing system.

Thematic Applications of IRS-Data

IRS 1-A and IRS 1-B (planned for 1991) provide data with finer spatial resolutions (72.5 m for LISS-I and 36.25 m for LISS-II) and has spectral channels close to that of LANDSAT-TM, thus ensuring continuity of remote sensing data for users. The repetivity cycle of 22 days and the radiometric sensitivity of 128 quantisation levels of IRS-1 A (and IRS-1 B) are considered optimum for most of the applications. Various thematic application projects have been taken up under IRS utilisation programme to operationalise remote sensing technology for the National Natural Resources Management System (NNRMS). The application areas wherein different spectral bands can be used are given in Table 9.2. Some of the application projects carried out on operational basis using IRS-1A data are mentioned in the subsequent paragraphs.

TABLE 9.2: Theme Specific Applications of IRS-1A Data

Band	Spectral range (Microns)		Application
1.	0.45 - 0.52	(i)	Coastal environment studies (Coastal Morphology and sedimentation studies)
		(ii)	Soil/vegetation differentiation
		(iii)	Coniferous/Deciduous vegetation discrimination
2.	0.52 - 0.59	(i)	Vegetation vigour
		(ii)	Rock/soil discrimination
		(iii)	Turbidity and bathymetry in shallow waters.
3.	0.62 - 0.68	(i)	Strong chlorophyl absorption leading to discrimination of plant species.
4.	0.77 - 0.86	(i)	Delineation of water features
		(ii)	Landform/geomorphic studies.

Remote sensing inputs especially from LISS-II of IRS-1A were provided to the technology mission on drinking water and related water management for scientific source finding. Ground water potential maps on 1:250,000 scale have been prepared for several states in the country, and 1:50,000 scale maps were prepared for selected districts after detailed investigation, which will be extended for the rest of the country in future.

Landuse/land cover mapping for few districts using visual and digital technologies, have been developed and the same is being taken up on 1:250,000 scale wasteland mapping, and wasteland delineation for some of the critically affected districts at 1:50,000 scale has been done on nation wide mode. IRS-1A data is being used to update these maps.

Monitoring of forest areas under three different categories of closed, degraded and open was done using IRS data.

Seasonal inventory of surface water bodies and reservoir water quality studies were carried out using IRS-1A data. Mapping of flood affected areas in Ganga and Brahmaputra river systems and snow covered areas in Beas and Chenab catchments are other applications of IRS data.

The Department of Agricultural and Co-operation and the Department of Space have launched a joint collaborative project for crop acreage and production estimation using IRS data. Crop stress detection and crop yield modelling are at technique development stage. Further priority delineation of some of the watersheds for soil conservation measures has been attempted using sediment yield index concept. Continuous monitoring and study of the environmental changes like decrease in vegetation cover, impact of mining are envisaged using IRS-1A data.

Satellite data has unique characteristics such as uniformity of view over very large area under constant illumination conditions, near orthogonality, better definition of certain features in some of the specific bands and digital data which enable it to be used for thematic mapping. The improved spatial resolution of IRS-1A and relief information from next generation IRS satellites (similar to SPOT) may make it possible to produce topographic maps on larger scales.

Application Potential of IRS-IA Data

Some of the salient features of the demonstrated potential of IRS-1A data that emerged in the National Seminar on "Indian Remote Sensing Satellite (IRS-1A) mission and its application potential" organised by Department of space at NRSA, Hyderabad during 21-22 Dec. 1988 are listed below:

1. *Cartography*: 1:250,000 scale maps could be adequately updated using IRS-1A data.
2. *Land use*: IRS-1A data is comparable to landsat data. Land cover boundaries were much sharper in LISS-1 data from IRS-1A compared to MSS data from LANDSAT.
3. *Forestry*: LISS-1 data could easily replace LANDSAT MSS data while LISS-II could be used to supplement LANDSAT TM data in mapping forest cover.
4. *Hydrology*: Designated Croplands and small water bodies could be delineated using LISS-1 data. Qualitative mapping of turbidity levels in reservoirs was possible.
5. *Geosciences*: IRS-1A data was found better compared to LANDSAT for litho-structural mapping. However, in hilly areas, IRS-1A data does not show much variations, probably due to high sun elevation angle compared to LANDSAT. IRS-1A data can be effectively used for deriving geology and geomorphology related details.
6. *Soils*: Utility of IRS-1A was clearly demonstrated for small scale mapping of soils.
7. *Coastal geomorphology and Oceanography*: IRS-1A data could be utilised in place of LANDSAT for coastal wetlands mapping and land form studies.

Conclusion

Use of IRS-1A is in consistence with the national goal of diffusing the benefits of new and emerging high technology systems with the mainstream of the country's developmental efforts, and recognising the potentials of space based remote sensing as a powerful technique for the efficient and optimal management of country's natural resources (I.S.R.O., 1989). The availability of four band data in visible and infrared regions and having spatial resolution of 36m and 72m with a periodicity of 22 days has opened up new vistas for applications in monitoring and management of the resources and the environment. The quality of data is comparable to the contemporary international satellites while cost of data is much lesser.

The IRS-1A data can be manipulated and processed in several ways by digital image processing techniques for bringing out different themes of national and regional relevance. The information thus generated can be integrated with other sources for the preparation of the thematic maps. It can also be a valuable input to the development of the Geographical Information System. A single data set can be used for generating information for several themes with appropriate processing and experimentation.

REFERENCES

I.S.R.O. (1986), *The Indian Remote Sensing Satellite System IRS*, Indian Space Research Organisation, Bangalore.

I.S.R.O (1989), *National Seminar on Indian Remote Sensing Satellite — IA, Mission and its Application, Potentials: Conclusions and Recommendations,* (December 21-22, 1968, Hyderabad). Indian Space Research Organisation, Bangalore.

N.R.S.A. (1988), *IRS Data Users Handbook*, National Remote Sensing Agency, Document No. IRS/NRSA/NDC/HB/01/86, Hyderabad.

10

ROLE OF AERIAL PHOTOGRAPHY AND SATELLITE REMOTE SENSING IN THEMATIC CARTOGRAPHY: A Case of Forest Mapping

A.K. TIWARI

It is well known that the increasing population growth and developmental activities have caused heavy pressure on the forests, and, we are well aware that this renewable resource will not remain inexhaustive unless a balanced approach to harvest versus preservation is developed. Keeping in mind the conservation, a forest manager needs up-to-date and accurate information of various forest related parameters.

A variety of remote sensing data and techniques have been made available for the mapping of the forests and related studies during the last few years which, when used effectively are able to furnish the forest managers much more information in shorter time and at a lower cost but with a greater accuracy.

The aim of the present chapter is to describe the up-to-date forest information extraction possibilities from various types of remote sensing data for forest mapping.

Remote Sensing Data and Techniques

Remote sensing data used in forestry can be classified into (a) aerial and (b) satellite data; based on platform of the sensor. Aerial remote sensing data include aerial photographs as well as air borne multispectral data. A number of satellites e.g. LANDSAT, SPOT, IRS etc. provide multispectral remote sensing data which can be used for forest mapping.

Information from aerial photographs can be used through visual interpretation using various instruments (Tomar and Maslekar, 1974) whereas multispectral data (airborne or satellite) can be interpreted visually or with the aid of computers DIP (Digital Image Processing). For visual interpretation of multispectral data the False Colour Composite (FCC) pictures are generated by superimposition of images for green, red and infrared wave length bands. Digital image processing has been proved to be very useful in extracting forest related information through various enhancements and classification algorithms.

Demonstrated Applications

Forest Cover Type Mapping

The aerial photographs are being used for forest type mapping for about four decades (Boon, 1956, Stellingwerf, 1959). The aerial photographs when viewed through stereoscopic instruments provide the three dimensional view of the objects which enables the interpreter to measure the dimensional parameters of trees, i.e. crown width, crown depth and height. The identification of forest species mainly depend on the scale of aerial photographs, variations in form and strature of species, in addition to skill and experience of the interpreter.

In India a number of workers and organisations have used aerial photographs for forest mapping. Forest Survey of India has been using black and while aerial photographs for preparation of 1:50,000/63,350 scale maps since mid 1960s. Tomar and Maslekar (1974) based on their long experience in forestry related studies suggested a classification system for forest survey and mapping of Indian forests with special reference to photo interpretation. Shedha (1979) studied the Gudem Reserve Forest of Kagaz Nagar Forest Division of Andhra Pradesh by aerial photo interpretation as well as by ground methods. He compared the accuracy, time and cost of the survey by these two methods and concluded that photo interpretation technique is more accurate, less time consuming and less expensive than ground method.

Tiwari (1978) made a comparative evaluation of forest cover types and landuse classification and mapping with aerial photographs vis-a-vis conventional ground mapping in Tehri Garhwal in Himalayas and reported that a more accurate landuse and forest map could be prepared using aerial photographs in lesser time and at lower cost.

Tiwari (1983) and Tiwari and Singh (1987) mapped different forest types in Kumaun Himalaya on the basis of Importance Value Index (IVI).

An example of use of air-borne multispectral data in forest type mapping in Godavary Basin. In this study he classified teak plantations, anogeissius forests, forests, mixed forests — poor stocked, mixed forests — well stocked, mixed forest medium stocked, degraded forests etc.

A number of forest mapping studies have been carried out using satellite data through visual as well as computer aided classification. Visual interpretation was mainly used for mapping of forests according to canopy closure viz. open forests, closed forests etc. (Madhavan Unni, 1983). Roy (1982) suggested that identification of different forest cover types can be done using the images of different seasons and superimposition of interpreted maps.

Supervised classification has mainly been used for forest type classification through digital image processing. The studies carried out by various workers have indicated that major forest types e.g. tropical ever green, tropical semi ever green, tropical deciduous etc. can be delineated through supervised classification of landsat data. Using the data of leaf fall season presence of under storey bamboo in part of Nallamalai Hill Range, was successfully delineated (Madhavan Unni, 1983).

Singh (1987) carefully examined the separability of different tropical forest cover classes using LANDSAT MSS data. He concluded that forests can be classified on the basis of canopy closure as the canopy closure cause significant difference in spectral response in LANDSAT MSS data. However, bare soil, grassland and regrowth from cultivation and regrowth from open forest cannot be separated accurately. Mukai *et. al.* (1980) and Ohnuki (1987) has suggested that accurate forest type classification can be performed using LANDSAT MSS data if the data of two seasons are merged to be used as input subscene.

The Thematic Mapper (TM) of Landsat -4 and -5 produces an order of magnitude more data per scene than the LANDSAT multispectral scanner (MSS). The TM has more spectral bands and greater radiometric resolution than the MSS. The six reflective TM bands include a blue band (TM Band 1) and two short wave infrared bands (TM Bands 5 and 7) in wave length ranges not sensed by MSS. A number of studies using simulated TM data revealed that TM data offer significant improvements over MSS in classification accuracy and vegetation cover types. A classification experiment using actual TM data has confirmed the simulation studies by showing that the TM permits more detailed and accurate classification that the MSS (Ahren, 1987).

Timber Volume Inventory

Aerial photographs have played a significant role in timber volume inventory. The inventory involves relationship of ground measured timber volume to the photo parameters like crown cover, tree stand height etc. The timber volume inventory in ground is done mainly through existing volume table. The volume tables for the ground are based on the regression analysis between tree parameters viz. girth, height, tree density and the timber volume per tree or per unit area.

The timber volume inventory or growing stock estimation using aerial photographs is based on the relationship between photo parameters and ground parameters. The relationship is used for preparation of aerial stand volume tables. In the past the construction of aerial stand volume tables was regarded as the domain of research in same manner as preparation of yield or volume tables for ground measurements (Tiwari, 1982). In India, a number of workers have used the techniques for volume estimations in working plan preparation and other studies (Tomar, 1976, Shedha, 1983. Tiwari (1982) reviewed the work done by more than thirty Indian workers in different parts of the country.

Traditionally medium to large scale photographs were used for volume estimations. Small scale photography is often regarded by foresters as inadequate. Marshall and Meyer (1977) in Mannesota found that 1:24,000 scale photos to be marginally acceptable at best. Aldrich (1979) suggested that resolutions needed for such studies required atleast 1:20,000 scale photos. To deal with small scale photos, Maclean (1981) have evaluation double sampling method for stratification, a design well suited to regional timber inventories (Bickford, 1952). The design as described by Bickford (1952), consists of an intensive grid of photo plots which are sorted into photo volume classes and sub-sampled on the ground. The number of plots in a given photo class determine the class area and the ground measurements determine the mean volume per acre associated with that class. Maclean found that the double sampling design provided total timber volume with twice the efficiency (the amount of information per unit of cost) of a simple field plot survey i.e. precision was improved without an increase in cost. Results were similar to those obtained from the earlier studies testing interpretation on 1:20,000 scale photos.

Multispectral data can be used for Timber Volume Inventory through digital image processing techniques by two methods:-

(a) *Through Stratification*

In this method the large scale aerial photographs are stratified in forest type, crown cover and height. For each strata field inventories are made for the volume. Selected strata are used as ground truth for the supervised classification of LANDSAT scene. By this method Ohmuki (1987) reported about 76 per cent accuracy.

(b) *Through Regression Analysis*

This is based on the spectral relationship between spectral response and timber volume. In this method Mukai *et al.* (1980) used band 5 and 7 of LANDSAT MSS data of two seasons. The timber volume for sample sites was related to LANDSAT MSS spectral response over the sample sites through multiple regression:

$$V = 30.5\, x_1 - 16\,2\, x_3 - 9.1\, x_4 + 955.5$$

where,

V = timber volume M_3/ha $\quad$ $x_1 = Ba_5$ of February

$x_2 = B_7$ of February $\quad$ $x_4 = Ba_7$ of March.

Similarly, they developed relationship between ground measured productivity (yearly growth) and spectral response through the equation

$$G = -0.30\, x_1 - 0.16\, x_2 - 0.60\, x_3 + 0.15\, x_4 + 16.4$$

where,

$x_1 = Ba$ 5 Feb, 1973; $\quad$ $x_2 = Ba$ 7 Feb, 1973

$x_3 = Ba$ 5 Mar, 1973; $\quad$ $x_4 = Ba$ 7 Mar, 1979

PLATE 10.1: Supervised Classification of Landsat MSS

Assessment and Mapping of Forest Damage

Shedha (1982) used colour infrared and multispectral aerial photography on 1:10,000 scale for detection of diseased and dead vegetation around Dehradun. This study revealed that healthy teak, sal, eucalyptus and trees of other species appear in bright reddish hues. Diseased and dead trees are registered in bluish green and defoliated trees in greenish blue or false colour aerial photographs. Band -1 (blue of multispectral aerial photographs exhibited diseased Erythrina trees and dead bamboo clumps in light gray tone).

The image substraction method is the most common tool for forest damage identification. All kinds of forest depletions can be monitored through this technique. Ohnuki (1987) has reported that in Japan the quantification of diseased damage caused by pine nematode had been successful with 100 per cent accuracy. The estimation was done through the substraction of recent LANDSAT image bands from the old LANDSAT image bands, when disease was not present.

Nelson and Holben (1987) carried out investigations in Brazil using multi resolution satellite data. In this study LANDSAT MSS AVHRR —Local Area Coverage (LAC), AVHRR—Global Area Coverage (GAC) and Geostationary Operational Environmental Satellite (GOES) data were processed to differentiate forest from non-forest. The results of MSS were compared with those of LAC, GAC and GOES. It was found that:

(a) LAC data are capable of adequately delineating colonization clearing,
(b) The spatial resolution of the GAC data is too large to delineate linear clearings of varying length (tens to hundreds of kilometres) up to 2 km. wide reliability,
(c) The visible GOES data were to little utility due to excessive data noise, and
(d) Probability thresholding procedures discriminated forest from non-forest more accurately than vegetation index thresholding procedure.

Vogelmann and Rock (1986) assessed the decline in coniferous forest using TM stimulator data. Using field instruments they decided the band ratios to discriminate low and high damage sites using TM simulator data 1.65/1.23 and 1.65/0.83 μm band ratios were found to be strongly correlated with ground based measurements of forest damage. Ratio colour density slice images using 1.65/1.23 or 1.65/0.83 μm band ratios, and images using 0.56 μm and 1.65 μm bands with either of these band ratios in false colour composite provided accurate means of detecting, quantifying and mapping levels of forest declines.

Site Quality Assessment

The assessment of forest site quality is becoming increasingly important at local and regional levels, the data are needed for detailed management of forest tract at regional level. Shedha and Kumar (1980) reported that interpretation of qualitative attributes of forest vegetation along with the measurements of quantitative parameters on 1:10,000 scale aerial photography have proved to be useful in site quality assessment of Sal Forests of Dehradun. Bonnor and Morrier (1981) conducted studies on site quality assessment using 1:20,000 scale aerial photographs on forests near Pembroke, Ontario in Ottawa Valley. Through comparison of photo assessment and ground estimates, 76 per cent accuracy was found. It was concluded by Bonnor and Morrier (1981) that the accuracy can improve if the interpreter becomes more familiar with the procedure and the inventory area.

Afforestation Site Selection

Successful use of aerial photographs (1:10,000) scale in evaluation of land suitability and site selection for afforestation of suitable species were demonstrated by Pandey and Shedha (1981) through a case study in Sauni Forest of West Almora Forest Division of Uttar Pradesh. Potential afforestation sites in the area have been mapped on 1:25,000 scale and suitable species for fuel, fodder and timber have been suggested for each site. Sharma and Negi (1981), prepared a map identifying sites for social forestry in part of Balaghat District of Madhya Pradesh with the help of 1:50,000 aerial photographs. Prasad *et al.* (1980) conducted soil survey and prepared soil map using 1:50,000 scale aerial photographs and evaluated their suitability for raising teak plantation in part of Bilaspur District of Madhya Pradesh. The suitability classes used were highly suitable, moderately suitable, marginally suitable and non-suitable. Seth and Tomar (1973) used aerial photographs on 1:60,000 scale for forest resources survey over 10,416 km^2 in East Godavari, Andhra Pradesh. The photographs were used for location of sample plots on ground eliminating non-forested plots from ground sampling and for landuse and forest cover type map preparation. The forest land was further stratified in (a) high volume, (b) medium volume, (c) low volume, and (d) plantation. The three volume strata were further sub-classified into five crown density classes of 20 per cent. Suitable plantation sites were also mapped.

Conclusion

The variety of remote sensing data and techniques provide a great scope for forest related study. Aerial remote sensing can be applied effectively in preparation of large scale forest maps showing important species composition, stand height and crown density classes. Maps showing available afforestation sites can also be prepared. Large scale photographs have greater scope in preparation of timber volume inventories and assessment of site qualities. The regional estimates of timber volume can be done through digital satellite data. False colour and multispectral aerial photographs have proved very useful in separating dead, diseased and healthy trees. Remote sensing techniques have proved to be very useful in periodical monitoring of forest vegetation. Both, aerial and satellite remote sensing can be used for the monitoring purposes. The maps generated through visual interpretation of aerial photographs or satellite images for two different periods can be compared for broad forest cover changes. The changes like disease damage etc. can be monitored through digital satellite data or air borne data for two different periods.

REFERENCES

Aldrich, R.C. (1979), Sampling for multi-resource information. Can aerial photographs help and at what cost? in W.E. Frayer (ed), *Proceedings of the Workshop on Forest Resource Inventories*, Colorado State University, Forest Collins, pp. 393-405.

Bickford, C.A. (1952), The sampling design used in forest survey of the north-east, *Journal of Forestry*, vol. 50, pp. 290-3.

Bonner, C.M. and K.C. Morrier (1981), Site classification from air photos in a forest inventory, *The Forest Chronclie*, vol. 57, No. 6, pp. 265-7.

Boon, D.A. (1956), Recent developments in tropical forest, *Photogrammetrica*, vol. 12, pp. 382-6.

Horler, D.N.H. and Ahern (1986), Forest information contents of Thematic Mapper data, *International Journal on Remote Sensing*, vol. 7, pp. 405-28.

Maclean, C.D. (1981), Timber volume stratification on small scale aerial photographs, *Journal of Forestry*, vol. 79, pp. 739-40.

Madhavan Unni, N.V. (1983), Application of remote sensing in forest mapping and management with special reference to air borne and multispectral techniques, *Proceedings of the NNRMS Seminar*, Hyderabad.

Marshall, J. and M. Mayer (1977), A field evaluation of small scale forest resource aerial photography, *LAFHE RSL Research Report 772*, Remote Sensing Laboratory of the College of Forestry and the Agriculture Experiment Station, Institute of Agriculture, Forestry and Home Economics, K.P. University, St. Paul.

Mukai, Y., S. Tanaka, S. Tekeuchi, I. Nekajima, I. Ohnoki, H. Wantanabe and R. Hatamura (1980), Application of Landsat data for the management of forest resources, Commission No. VII, W.C. VII/1, Japan.

Nelson, R. and B. Holben (1987), Identifying deforestation in Brazil using multi-resolution satellite data, *International Journal on Remote Sensing*, vol. 8, pp. 429-48.

Ohnuki, I. (1987), Application of satellite remote sensing to forestry, Lecture note, Forestry and Forest Products Research Institute, JICA, Japan.

Pande, L.M. and M.D. Shedha (1981), Land suitability and site selection for afforestation suitable species using aerial photographs : A case study of Ranikhet area of Kumaun Region, Lower Himalaya, in the *Proceedings of the Workshop on Modern Techniques for Site Identification for Afforestation and Pasture Development*, Indian Photointerpretation Institute, Dehradun.

Prasad, J. and M.L. Manchanda (1981), Site evaluation for teak plantation : A Case study of a part of Bilaspur District, M.P., *Proceedings of the Workshop on Modern Techniques of Site Identification for Afforestation and Pasture Development*, Indian Photointerpretation Institute, Dehradun.

Roy, P.S. (1982), Vegetation mapping and forest conditions monitoring through visual analysis of multidate Landsat data, presented at the National Conference on Application of Remote Sensing Data in Natural Resources, Environment, Landuse and Problems relating to Training and Education.

Seth, V.K. and M.S. Tomar (1973), Contribution of small scale photography in forest resource survey of east Godavari, *Indian Forestry*, vol. 99, No. 2.

Sharma, M.K. and G.S. Negi (1981), Role of aerial photographs in site identification for social forestry, *Proceedings of the Workshop on Modern Techniques of Site Identification for Afforestation and Pasture Development*, Indian Photointerpretation Institute, Dehradun.

Shedha, M.D. (1979), Forest mapping by ground and photo methods: A case study, *Van Vigyan*, Vol. 19 (1979-81).

——— (1982), Forest tree disease and insect pest detection with the help of multispectral and colour infrared photographs, *Proceedings of the Natural Science Symposium*, vol. 3, pp. 448-9.

——— (1983), Application of remote sensing in forest mapping and management with special reference to aerial photointerpretation, *Proceedings of the NNRMS Symposium*, Hyderabad.

Shedha, M.D. and S. Kumar (1980), Forest site quality assessment by aerial photographs, unpublished report, Indian Photointerpretation Institute, Dehradun.

Singh, A. (1987), Spectral separability of tropical forest cover classes, *International Journal on Remote Sensing*, vol. 8, pp. 97-9.

Tellet, P.M., B. Guidon and D.G. Goodenough (1981), Forest classification using simulated Landsat Thematic Mapper data, *Canadian Journal of Remote Sensing*, vol. 7.

Tiwari, A.K. (1982), Forest cover type mapping and assessment of growing stock of part of Amarabad Plateau, Andhra Pradesh, Training Report, Indian Photointerpretation Institute, Dehradun.

Tiwari, A.K. and J.S. Singh (1987), Analysis of forest landuse and vegetation in part of Central Himalayas using aerial photographs, *Environmental Conservation*, vol. 14, pp. 233-44.

Tomar, M.S. (1976), Use of aerial photograph in working plans, *Indian Forester*, vol. 102 (2) pp. 98-108.

Tomar, M.S. and A.R. Maslekar (1974), *Aerial Photographs in Landuse and Forest Surveys*, Jugal Kishore & Co., Dehradun.

Tiwari, K.P. (1978), A comparative evaluation of landuse and forest type classification and mapping with aerial photographs vis-a-vis conventional ground stock mapping : A case study in Tehri Garhwal Himalaya, *Indian Forester*, vol. 104, pp. 757-67.

Vogelmann, J.E. and B.N. Rock (1986), Assessing forest decline in coniferous forests of Vermant using NS-001 Thematic Mapper Simulator data, *International Journal of Remote Sensing*, vol. 7(10), pp. 1303-21.

Willingham, J.W. (1957), Estimation of forest management inventory data from aerial photographs, *Photogrammetric Engineering*, Vol. 8, pp. 203-15.

11

SOFTWARE OPTIONS FOR THEMATIC INTERPRETATION AND MAPPING USING REMOTELY SENSED DIGITAL DATA

S.K SAHA AND M. KUDRAT

Remote sensing technology has opened a new dimension for collecting real time data about resources in a short time covering larger areas at periodic intervals. In addition to this recent advances in efficient computer hardwares and softwares for image processing of remotely sensed digital data have opened new vistas in interpreting, mapping and monitoring of natural resources. Digital image processing involves the manipulation and interpretation of digital images through various softwares with the aid of a computer. Today, the sources of digital remote sensing data used for thematic mapping are many — these sources range from earth resource satellite systems (e.g., LANDSAT–TM, SPOT, IRS-1A), to meteorological satellites (e.g., NOAA, INSAT, GOES); to air borne scanner data, to digitized aerial photographs generated by scanning microdensitometer.

Softwares Used for Digital Image Processings of Data

Computer software based image processing methods for analysis of digital data are literally infinite. However, virtually all these methods may be categorized into following types of computer aided operations:

(a) Image rectification
(b) Image enhancement
(c) Image classification
(d) Digital change detection
(e) Data integration using GIS (Geo-Information System) softwares)

The selection of above softwares based digital analysis techniques depends upon the forms of digital raw data from various remote sensing sensors, the type of information the user is hoping to extract from the data and the desired from the output.

(a) *Image Rectification*

This digital operation acts to *restore* geometrically distorted image data to a more faithful representation of the original scene. It involves correcting for variety of geometric distortions due to scanner distortion, altitude and attitude (roll, pitch and yaw), variations of space craft, that may be present in the original image data. The correction process involves use of Ground Control Points (GCP) which are features like intersections of roads, confluences of rivers, small water bodies etc., of known ground location that can be accurately located on image. In the correction process numerous GCPs are located in terms of both their image co-ordinates (scanline, pixel) and map co-ordinates (latitude, longitude). Differences between the observed and actual GCP locations are used to determine coefficients using a least square regression analysis, for two transformation equation that relate the map and image co-ordinates:

$$u = f(x, y) \qquad (1)$$
$$v = g(x, y) \qquad (2)$$

where, (u, v) = image co-ordinates;
(x, y) = map co-ordinates;
(f, g) = transformation equation.

The corrections process consists of the following steps:

(i) Define a geometrically uniform *output* matrix in terms of the map co-ordinates,
(ii) Using the equations (1) and (2) transform the map co-ordinates of each cell in the output matrix into the corresponding image coordinates,
(iii) Transfer the pixel gray values at this image co-ordinates into the cell in the output matrix (Fig. 11.1). This procedure of geometric transformation of image is called resampling. The resampling methods commonly available are nearest neighbour, bilinear

interpolation and cubic convolution. Details of these techniques are given in Table 11.1 (Curran, 1985).

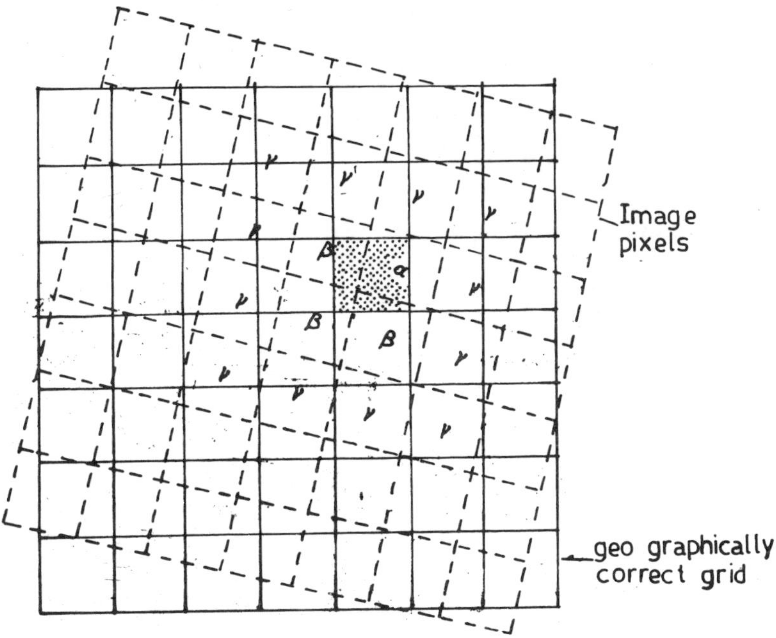

FIG. 11.1: Resampling procedures used to geometrically correct an image (*Source*: Curran, 1985).

As we will discuss later, resampling techniques are also important in several digital processing operations besides the geometric correction of raw images. For example, re-sampling is used to overlay or register multidate images. It is also used to register images of differing spatial resolutions. Finally, resampling procedures are used extensively to register image data and other sources of data in GIS.

(b) *Image Enhancement*

The apparent contrast between features in the image can be accentuated by using digital enhancement techniques. This greatly increases the amount of information that can be visually interpreted or subject to other digital analysis, from the raw image data. A wide variety of digital enhancement techniques softwares are available and they can generally be characterised as point or local operations. Point operations modify the values of each pixel in an image data set independently. Local operations modify the value of each pixel in the context of the pixel values surrounding it. Commonly used digital enhancement techniques can be categorized as follows:

i. Contrast manipulation
 — Contrast stretching
 — Density slicing
ii. Spatial feature manipulation
 — Edge enhancement
 — Spatial filtering
iii. Multi-image manipulation
 — Spectral band ratioing and differencing
 — Principal component and canonical component analysis
 — Spectral vegetation indices
 — Intensity-hue-saturation (IHS) color space transformation
iv. Auxiliary enhancements.

TABLE 11.1: **A comparison between three methods of resampling of remotely sensed images**

Method	Technique	Example pixels to be transferred in Fig. 11.1	Relative computing time	Advantage	Disadvantages
Nearest Neighbour	Transfer gray value of nearest pixel	α	1	Sample to compute, Gray values unaltered	Image disjointed pixel, offset by upto half a pixel
Bilinear Interpolation	Transfer proximity weighted average of nearest 4 pixels	α and β	10	Smooth image, geometrically accurate	Gray values altered
Cubic convolution	Transfer evaluated weight of nearest 16 pixels	α, β and γ	20	Very smooth image	Complex to compute, Gray values altered.

(*Source*: Curran, 1985)

i. Contrast manipulation

Contrast Stretching

The objective of contrast stretching is to expand the narrow range of gray values (DN Values) typically present in an input image over a wider range of gray values for producing an output image with the optimum contrast. Commonly applied contrast stretching methods are linear stretching and non-linear stretching. Some contrast stretch processes are illustrated in Fig. 11.2. In linear stretching the range of image gray values are uniformly expanded to fill the total range of display device (Fig. 11.2c). The non-linear stretching such as histogram equalized stretch is used when it is felt necessary to weight DN (gray) values by their frequency of occurrence (Fig. 11.2d).

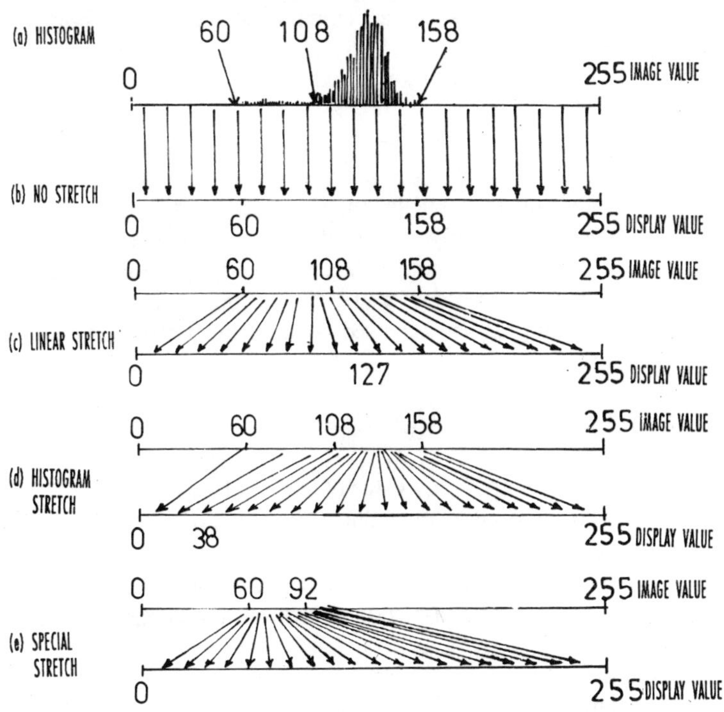

FIG. 11.2: Principle of some contrast stretch enhancements (*Source*: Lillesand and Kiefer, 1987)

Non-linear Gaussian stretch fits the original histogram to a normal distribution curve between 0 and 255 limits with the result that even the light and dark ranges are enhanced in contrast. Other types of non-linear stretches are logarithmic, cube root and arctangent.

It has to be noted that there is no single contrast enhancement technique which gives the optimum enhancement. One has to identify the elements of the scene that are of great interest for the particular application, then through inspection of the original histogram one can determine the optimum stretch needed.

Density Slicing

This is an enhancement technique whereby the almost continuous gray levels in an image are converted into a series of density intervals, or slices, each corresponding to a specified gray level range. Each sliced range is then displayed in a separate colour or with a separate symbol.

ii. Spatial Feature Manipulation

Edge Enhancement

This is a digital enhancement technique for emphasising subtle gray scale variations. Where gray values of adjacent pixels vary by more than a pre-determined threshold value, the interface is marked on the image display with a contour, a step in the grey scale or a colour change. One of the algorithms used to implement this technique is given by:

$$Q(\text{i}) = P(\text{i}) + [P(\text{i}) - K.A(\text{i})] \tag{3}$$

where, Q(i) = Output gray value of the *ith* pixel;
P(i) = Original gray value of *ith* pixel;
A(i) = Local average of the *ith* pixel by taking an $n \times m$ neighbourhood where n and m are odd integers and $k = a$ constant.

This operation makes a pixel that is brighter than its surrounding brighter yet and a relatively, dark pixel, darker.

Spatial Filtering

Spatial filtering provides a means of improving images by suppressing or enhancing certain spatial frequencies. Spatial frequency refers to the *roughness* of the tonal variations occuring in an image (Lillesand and Kiefer, 1987). Image areas of high spatial frequency are those areas where the gray values change abruptly over a relatively small number of pixels (e.g., linear geological features like joints, fractures and faults). In low spatial frequency image areas, the gray values vary only gradually over a relatively large number of pixels (e.g., large water bodies or agricultural fields). Spatial filtering methods basically categorized into high pass filtering and low pass filtering.

High pass filtering emphasizes the detailed high spatial frequency components and de-emphasize the more general low frequency information.

Low pass filters are designed to emphasize low spatial frequency features and de-emphasize the high frequency components of an image.

iii. Multi-image Manipulation

Spectral Band Ratioing and Differencing

These types of enhanced images are generated by ratioing or substracting the DN (gray) values of pixels in spectral band with the corresponding pixels in another spectral band and contrast stretching the result. This ratio or difference image will suppress detail common to the two images and will enhance detail that is different between two images (Townshend, 1981). Ratio images are free of the effects due to variations in illumination across a scene. The number of possible ratios that can be developed from '*n*' bands of data is $n(n-1)$. Colour composites can be prepared using any three ratio images or using a combination of two ratio images and any individual band image.

Principal Component (PC) and Canonical Component (CC) Analysis

These multivariate analysis techniques used to reduce the number of spectral components (bands) to fewer principal or canonical components (few orthogonal axes), which account to most of the variance in the original multi-spectral images. Principal components are simply linear combinations of gray values of original bands associated with a co-efficient. For multispectral data, the first principal component (PC) includes the largest percentage of the total scene variance and succeeding components (PC2, PC3, ..., PCn) each contain a decreasing percentage of the scene variance. To avoid loss of information, the PC1, PC2 and PC3 can be combined to form False Colour Composite (FCC) for better analysis.

Cononical analysis, like PC analysis, compresses the information component of several images into one image and this maximises the difference in gray value between the major scene components (Curran, 1985). PC and cononical methods are very useful in analysing more number of bands data (like 7 bands of Landsat-TM), multi-temporal and integrated multi sensors data analysis.

Spectral Vegetation Indices (SVI)

SVI enhanced images are derived from empirical transformations of spectral responses in different spectral bands. These indices are useful for vegetation, agricultural crop features analysis. Commonly used SVI are —

$$\text{Ratio index (RI)} = \frac{\text{NIR}}{\text{Red}}$$

where, NIR-Reflectance in Near-IR band;
Red-Reflectance in red band

Normalized difference vegetation index (NDVI)
= (NIR - Red)/(NIR + Red)

Difference vegetation index (DVI) = (2 × NIR-Red)

Transformed vegetation index (TVI) = NDVI + 0.5

Perpendicular vegetation index (PVI) =
$(Reds - Redv)^2 + (NIRs - NIRv)^2$

where, Reds, NIRs – Reflectance of soil in red and Near-IR bands respectively;
Redv, NIRv – Reflectance of vegetation in Red and Near-IR bands respectively.

n – space spectral indices.

For Landsat-TM

Soil Brightness Index (SBI) = 0.331 B1 + 0.331 B2 + 0.551 B3 + 0.425 B4 + 0.480 B5 + 0.252 B7

Greenness Index (GI) = 0.247 B1 + 0.162 B2 - 0.406 B2 + 0.854 B4 + 0.055 B5 - 0.117 B7

Wetness Index (WI) = 0.139 B1 + 0.224 B2 + 0.403 B3 + 0.251 B4 - 0.701 B5 - 0.457 B7

(Crist and Cicone, 1984)

where, B1, B2, ... are spectral responses in LANDSAT TM Band 1, Band 2, ... respectively. SVIs have been used successfully for vegetation monitoring and vegetation conditions assessment. SVIs have also been used for estimation of various agrophysical parameters like leaf area index, biomass, crop yield etc. (Tucker *et al.*, 1979; Holben *et al.*, 1980; Kamat *et al.*, 1985).

Intensity-Hue-Saturation (IHS) Colour Space Transformation

Intensity relates to the total brightness of a colour, *Hue* refers to the dominant wavelength of light contributing to a colour. *Saturation* specifies the purity of a colour relative to gray. Digital images are typically displayed as additive colour composites using the three primary colours; red, green and blue (RGB). Transformation of the RGB components into IHS components provide enhanced image of more colour control (Schowengerdt, 1983). IHS enhancement operation has the ability to vary each IHS component independently without affecting the others. The IHS approach can also be used to display spatially registered data of varying spatial resolution.

iv. Auxilliary Enhancements

The image data can be enhanced by adding auxilliary data to them. For example, multispectral data can be merged with other sensor data with higher spatial resolution. LANDSAT TM data has been merged with high spatial resolution X-band, HH polarisation SAR (Synthetic Aperture Radar) data and the combined produce showed more information than the original individual images alone. Another technique is to integrate digital elevation terrain (DET) data with multispectral image pixels to produce synthetic stereoscopic images. Here, using the elevation data, image pixels positions are offset in accordance with their elevations (Lillesand and Kiefer, 1987).

(c) Image Classification

Image classification basically involves the analysis of multispectral image data and the application of statistically based decision rules for automatically categorizing all pixels in an image into land cover classes or themes. These categorized data may then be used to produce thematic maps of the land cover/thematic class present in an image and or to produce summary statistics on the areas covered by each land cover/thematic class type. Image classification digitally performs by mainly two methods: supervised classification, and unsupervised classification

Supervised Classification

Three basic steps are involved in a typical supervised classification procedure : training stage — in this stage, the image analyst identifies representative training areas and develops a numerical description of the spectral attributes of each land cover/thematic class type of interest in the scene; classification stage — each pixel in the image data set categorized into the land cover class most closely resembles; output stage — the typical forms of output products are — thematic maps, tables of area statistics of various land cover or thematic classes and digital data files amenable to inclusion in a GIS (Geo-Information System). Fig. 11.3 summarizes the basic steps in supervised classification.

There are several mathematical classifiers used in supervised classification in classification stage. The most popular are — the minimum distance to mean classifier; the parallelopiped classifier; and the maximum likelihood classifier.

The minimum distance to means classifier: First compute the mean spectral value in each band of each class and then a pixel of unknown identity is classified by computing the distance between the value of the unknown pixel and each of class means. This is the most simple and cheapest but less accurate classifier.

Parallelopiped classifier: Here the range of each class is defined by highest and lowest gray values in each band and appear as a rectangular area in two dimensional scatter plot. An unknown pixel is classified according to the class range or decision region, in which it lies or as *unknown*, if it lies outside all regions.

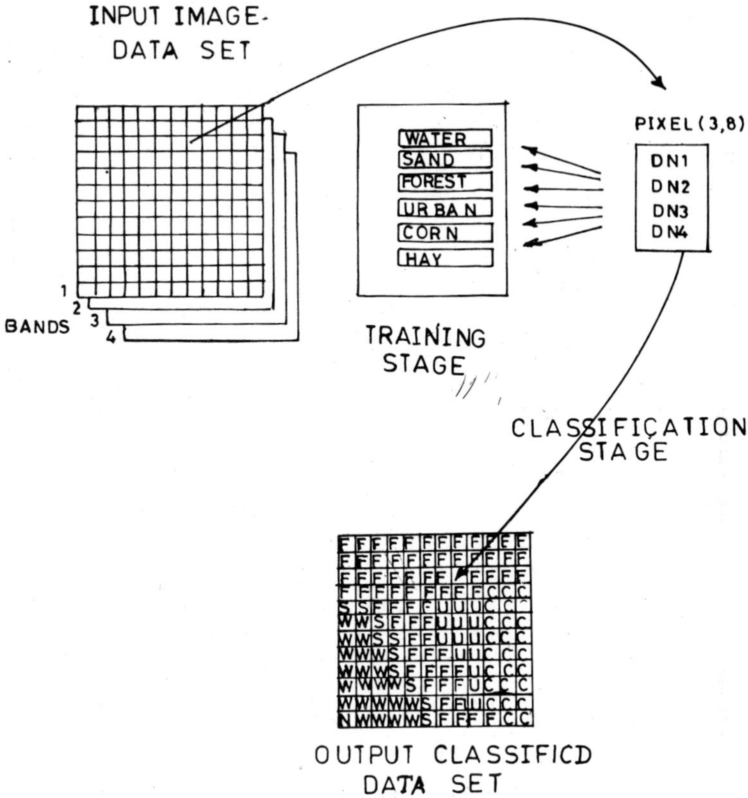

FIG. 11.3: Basic steps in digital supervised classification

Maximum likelihood classifier: This classifier quantitatively evaluates both the variance and covariance of the class spectral response pattern when classifying unknown pixel. This classifier computes the probability of unidentified pixel value belonging to each class and then assign it to the maximum likely class. If the probability values are all below a threshold specified by the analyst, the pixel is designated *un-classified.*

The last two classifiers in general provide more classification accuracy but with more processing time and cost.

Unsupervised Classification

In unsupervised approach the image data are first classified by aggregating them into the natural spectral groupings or cluster present in the scene using clustering algorithms. Then, the image analyst determines the land cover or thematic class identity of these spectral groups by comparing the classified image data to ground reference data. Unsupervised classification procedure computationally complex and quality of classification depends upon analysts knowledge about the land cover/thematic classes under analysis.

Multitemporal Image Classification

Various digital techniques are available for multitemporal image classification

Principal component transformation has been applied to reduce multitemporal data to fewer dimension and classification performed on first four principal components which contain as much as 95 per cent of the original information.

All available spectral bands data from various dates have been used for classification (supervised or unsupervised) without initial data compression.

The details of preprocessing, feature extraction and classification procedures for multitemporal image classification are presented in Fig. 11.4.

(d) **Digital Change Detection**

Digital change detection techniques require precise spatial registration and atmospheric corrections of digital images from multidates. A residual image is produced by various methods and then a threshold value, determined empirically or statistically, is applied in order to define areas of change. Presently various digital change detection techniques are available and the principal techniques are outlined below (Singh, 1986).

Image Differencing

In this technique, geometrically registered images of different dates are subtracted to produce a difference image which represent change between the two dates. Single band images or combinations of different bands, such as SVI (Spectral Vegetation Indices) can be employed in this technique. The empirical expression of this technique is given below.

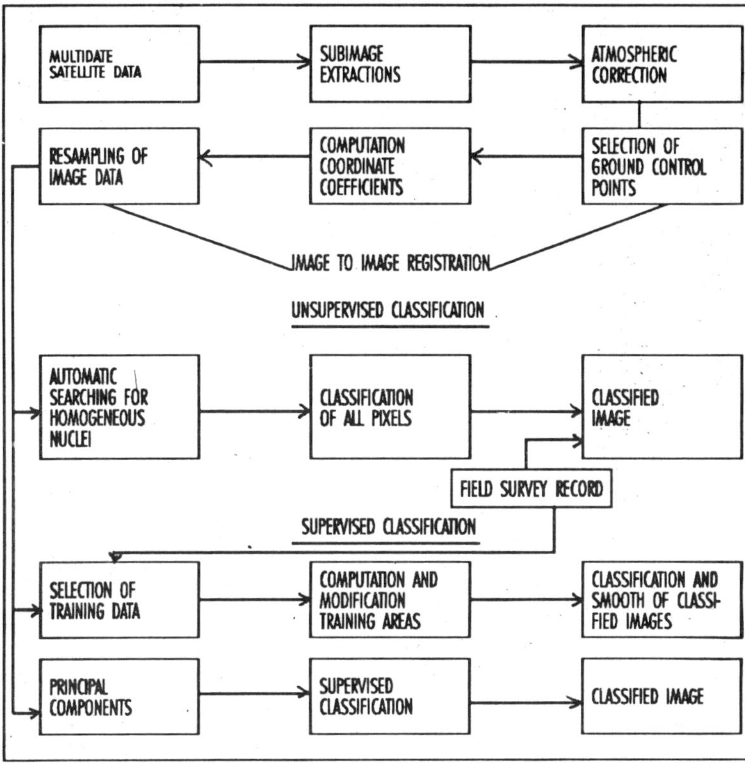

FIG. 11.4: Pre-processing, feature extraction and classification for multi-temporal satellite data.

$$Xijk = \frac{Xijk\,(t_2) - Xijk\,(t_1)}{2} + c$$

where, $Xijk$ = Difference in gray values of *ith* pixel with *jth* scanline *kth* band in two different dates

$Xijk\,(t_2)$ = Gray value of ith pixel with *jth* scanline *kth* band in second data t_2

$Xijk\ (t_1)=$ Gray value of *ith* pixel with *jth* scanline *kth* band in first date t_1

c = Constant used in order to generate output value 0 and 254.

Image Regression

In this method, it is assumed that pixel gray values at time t_1, $Xijk\ (t_1)$ are related to those at time t_2, $Xijk\ (t_2)$, by a linear function so that one image can be regressed against other using a least square method. An image differencing method is then applied that subtract the predicted values, $X\hat{i}jk\ (t_2)$, based on regression line, from the actual values in the second image, pixel by pixel (Ingram *et al*, 1981; Singh, 1986).

$$R\text{-}Xijk = \frac{Xijk\ (t_2) - Xijk\ (t_1)}{2} + c$$

This technique has the advantage of reducing atmospheric haze and sun angle effects.

Image Ratioing

Ratio change images are generated by dividing the pixels gray values in a particular band at time t_1, by the corresponding pixel gray values at time t_2. Pixel gray values higher or lower than one are taken to indicate areas of change.

Principal Component Differencing

In this change detection techniques, principal component analysis was performed first, separately for each data and principal component two or one (PC_2 or PC_1) from each data was used in the image differencing procedure. PC_1 and PC_2 were chosen because these components account for most of variance of the original data set.

Post Classification Comparison

This change detection method is a comparative analysis of spectral classifications for times t_1 and t_2, produced independently. This method requires more computer time than other methods.

Direct Multidate Classification

This approach is based on direct single time spectral classification either supervised or unsupervised, of a combined data sets of times t_1 and t_2, in order to identify areas of change (Singh, 1986).

(e) **Data Integration Using GIS**

Frequently, remote sensing data are readily merged with other sources of information in a GIS. For example, image data are often combined with soil, topographic, ownership, zoning and assessment information (Smith and Blackwell, 1980; Maher *et al.*, 1980; Maggio *et al.*, 1983).

GIS is a special type of information management system which provides for the entry, storing, manipulation, analysis, retrieval and display of spatially oriented data, through digital techniques. The major components of GIS are:

— Data inputting
— Data encoding
— Data manipulation
— Data conversation
— Data analysis
— Data management
— Data presentation

Schematic diagram showing the various components of a typical GIS is presented in Fig. 11.5.

Data Inputting

Typically the input data for GIS are—tabular data viz. socio-economic data; graphic data viz. thematic maps; field and laboratory data viz. physical and chemical data; digital data viz. remote sensing data — air-borne or satellite. Data inputting involves the conversion of lines, polygons etc. of thematic map data into digital formats like vector and raster using manual or semi-automatic or automatic digitizers.

Vector Format

In this system, the digital representation of lines and polygons information generally consists of an ordered sequence of *x-y* co-ordinates pairs.

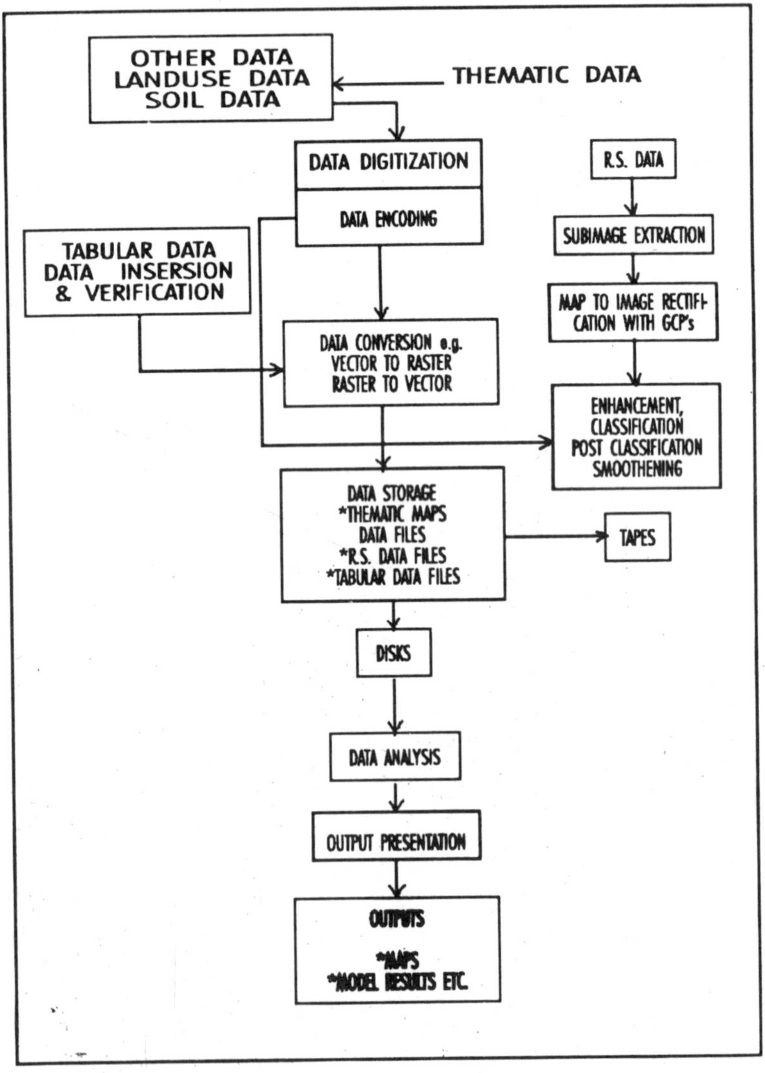

FIG. 11.5: **Schematic diagram showing the various components of GIS.**

Raster Format

This system consists of an array of matrix of grid cells — rectangular or squares.

Data Encoding

It involves the feeding of vector and raster data with locational codes into the data base. The one of the commonly used data encoding system is grid reference system based on latitude and longitude. The grid is most popular unit as it is cheap to store, manipulate and handle and it can be of any dimension.

Data Manipulation

It involves following operations by using softwares —

— Correction of remote sensing (RS) data with respect to topographical maps using rectification — resampling techniques based on Ground Control Points (GCPs).
— Post filtering smoothening of classified remote sensing data.

Data Conversion

It involves following operations —

— Conversion of vector data structure to raster form and *vice versa.*
— Combining tabular data with digitized data of corresponding areas.

Data Analysis

It includes various types of data analysis using softwares —

— Classification and digital analysis of remote sensing data to produce thematic maps;
— Digital change detection to monitor changes;
— Statistics of change detection;
— Areal extend computation;
— Modelling;

— Digital Terrain Modelling (DTM);
— Overlay of two or more data sets to create a composite data set.

Data Management

It basically involves data storage and retrieval to and from data base through relational data base management package and a higher level querry language which provides interface between user and data base.

Data Presentation

It involves following mentioned operations —

— Area tabulation, display and line printing of results of modelling, statistical analysis.
— Map generation using grid cell and polygon structured data.
— Display and printing of thematic maps.

There are several ways of incorporating remotely sensed data into GIS —

(a) Production of thematic maps through interpretation and analysis of R.S. data then digitization and incorporation,
(b) R.S. data is combined with other related geographical data such as Digital Terrain data to produce an improved single thematic map, and this map is then incorporated inside the data base,
(c) All data, including R.S. are placed in the data base prior to an interpretation or analysis. This enables to get an improved multispectral classification.

This method integrates several data prior to inclusion within the data base. It involves the sub-division of remote sensing data into photomorphic units which are areas marked by homogeneous appearance. Using other data planes such as soil, geological and vegetation maps, the photomorphic units are sub-divided to create integrated terrain units. Additional data planes containing features, such as administrative boundaries, water course lines etc., are registered with these terrain unit plain to create geo-data base.

Conclusions

The scope of digital image processing in the generic sense is virtually unlimited. With the increasing availability and sophistication of geographic information systems softwares and coupled with increasing availability of high resolution multispectral remote sensing data, the future of utilizing remote sensing data as an integral component of thematic interpretation and mapping, is indeed a bright one.

REFERENCES

Crist, E.P. and Cicone, R.C., (1984). Application of Tasseled cap concept to simulated Thematic Mapper data, *Photogrammetric Engineering and Remote Sensing*, Vol. 50(3), pp.343-52.

Curran, J.P., (1980). Image processing, *in* P.J. Curran (ed), *Principles of Remote Sensing*, Longman, London & New York, pp.176-226.

Holben, B.N.; Tucker, C.J. and Fan, Cheng-Jeng, (1980). Spectral assessment of soyabean leaf area and leaf biomass, *Photogrammetric Engineering and Remote Sensing*, Vol. 46(5), pp.651-6.

Ingram, K; Knapp, E. and Robinson, J.W., (981). *Change Detection Techniques Development for Improved Urbanised Area Delineation*. CSC/TM-81/6087, Report prepared for NASA, Computer Sciences Corporation, Springfield, Maryland.

Kamat, D.S.; Gopalan, A.K.S.; Ajai, Sashikumar, M.N.; Sinha, S.K.; Chaturvedi, G.S. and Singh, A.K., (1985). Assessment of water stress effects of crops, *International Journal of Remote Sensing*, Vol. 6(4), pp.577-89.

Lillesand, T.M. and Kiefer, R.W., (1987). Digital image processing, *in* Lillesand, T.M. and Kiefer, R.W. (eds), *Remote Sensing and Image Interpretation*, pp.610-99.

Maggio, R.C.; Baker, R.D. and Harris, M.K., (1983). A geographic data base for Texas Pecan, *Photogrammetric Engineering and Remote Sensing*, Vol. 49, pp.47-52.

Maher, R.V.; Boyd, M. and Longford, G., (1980). Integration of geographic and Landsat imagery: A case study of vegetation mapping in the white goat wilderness area, Alberta, *Canadian Journal of Remote Sensing*, Vol. 6, pp.86-92.

Schowengerdt, R.A., (1983). *Techniques for Image Processing and Classification*, Academic Press, New York.

Singh, A., (1986). Change detection in the tropical forest environment of Northeastern India using Landsat, *in* M.J. Eden and J.T. Parry (eds), *Remote Sensing and Tropical Land Management*, John Wiley and Sons Ltd., 237-54.

Smith, A.V. and Blackwell, R.J., (1980). Development of an information data base for watershed analysis. *Photogrammetric Engineering and Remote Sensing*, Vol. 46, pp.1027-38.

Tucker, C.J.; Elgin, J.H.; Mc.Murtrey III, J.E. (JR.) and Fan, C.J., (1979). Monitoring corn and Soyabean Crop development with hand held radiometer spectral data, *Remote Sensing of Environment*, Vol. 8, pp.237-48.

Townshend, J.R.G., (1981). Image analysis and interpretation for land resources survey, *in* Townshend, J.R.G. (ed.), *Terrain Analysis and Remote Sensing*. Allen and

12

PC AND CD – ROM SOLUTIONS FOR THEMATIC CARTOGRAPHY

PRITHVISH NAG AND JACK MASSEY

Thematic cartograpy has its own specialization and it continues to refine its techniques with the availability of data and ever-advancing technology. The huge data now available can not be ignored for such mapping. Experiments are continuously being done to produce data in such a form that it can easily be retrieved. The data on paper and cards were replaced by discs, floppies and computer compatible tapes. But, these remedies are also not considered the best way to store statistical data. Further, the printed maps are also considered voluminous. Atlas, a collection of thematic maps, will have several plates with useful and processed information. It becomes bulky for transportation and users trend to ignore the atlas however valuable maps it may contain. Furthermore, with the popularisation of the Geographical Information System (GIS), data handling and information processing including the preparation of maps has undergone revolutionary changes in their scope, coverage and associate technology. Obviously, the developments in electronics has influenced thematic cartography as well.

PC and CD-ROM

Personal Computers (PC) has come in big way in all walks of life. Thematic cartography is also not untouched by such advancements. Everyday, companies big and small, are bringingout solutions which are better more efficient and more versatile then the earlier ones. PC have proved to be cheaper and easy to maintain in comparison to mainframe or supermini computer systems. PC can do almost the same functions as the bigger systems. Wide range of PCs are available with varying memory capacities. It can be linked with more than one terminal or with the mainframe computer through local area network. PCs with compatible attachments can have a wide scope of applications including in Thematic Cartography.

The compact disc-read only memory (CD – ROM) technology has been developed a decade ago basically to record music. But, in last few years,- the same technology is being used for different applications, such as textual materials like reference book and manuals or dictionaries. A single disc can record about 550 megabytes of data which can rapidly be accessed by an IBM, PC AT/XT or a comparable PC based system. It also requires a CD – ROM reader, and of course a monitor. A single disc can store data equivalent to 250,000 4A size paper. The data is recorded permanently and can not be tampered. The information can be retrieved instantly and displayed pictorically in the form of thematic map as well.

A CD – ROM is a plastic and metallic, doughnut – shaped disc. The external diameter is 120 mm., while the internal diameter is 15 mm and it is 1.2 mm thick. A CD – ROM weights 18 gm. Data are stored in 2048 megabytes blocks on a spiral track stouting at the internal diameter. However, because of the use of what has become known as the *High Sierra Standard*, the data appear to the user as if they were stored in a typical MS-DOS file. Between 550 to 600 megabytes of data can now be stored on a single CD-ROM. Once the data are stored, they may not be altered, no case more data be added, hence the term, ROM, i.e., read-only-memory (Fig. 12.1).

Advantages and Limitations

The biggest advantage of this technology is the mass storage of data in digital form. It is associated with personal computers which has come in a big way recently. Large storage spaces can be saved in libraries, data centres, mapping institutions and the like. Mass production of CD-ROM discs are also cheaper and maintenance is easy. A single disc will cost from US 5-10 dollars. It cannot be easily damaged by scratching or dust, hence suitable in developing countries situation.

The other advantage of this system is that it can record maps as well. With appropriate software maps can be experimented and generated. Linked with a plotter, such a system can produce hard copy of maps at a various scales and projections. Further, with colour photo copier, a limited demand for large paper maps can be met. The maps are recorded in the vector form, such as DIME of the U.S. Bureau of Census (Cooke, 1987).

FIG. 12.1: PC and CD-ROM.

However, this system has its limitations as well. The maximum display of maps depends on the size of the colour monitor. It requires a large hard disc memory which may be expensive in developing countries. A CD-ROM leader is another expensive item which may not be easily available everywhere. Certain minimum environmental and maintenance aspects are also involved. In total, the whole system is associated with high technology and linked with the international market economics. The biggest overriding problem is the change of data in given time interval. In order to change over from conventional data or map in printed hard copies or on video discs or even on CCTs, an regular system of production of CD-ROM has to be assured.

It is apparent that the collection and processing of the statistical data bases of India have involved the protracted efforts of many persons and the expenditure of a considerable amount of money. This is an extraordinary investment in the creation of statistical resources.

Because of their ease of use and relatively low cost, the joint application of CD-ROM and PC technologies will help in exploiting this resource by making it available and useable by planners, researchers, teachers and students in India and around the world.

Applications of CD-ROM Technology

Compact disc–read only memory technology has a varied use in thematic cartography. In fact, this technology has made thematic cartography more people oriented and closer to the day to day needs of the masses. It has already commercialized thematic cartography. The rapid accessibility of the stored information is most important aspect of this technology.

In a country, like India high quality statistical data bases are available but the means to disseminate and use them effectively are not always possible. Attempts to solve this problem has made through the joint application of CD-ROM and PC technologies. Data can be permanently stored on a CD-ROM which can be installed in the CD-ROM drive attached to a PC and the appropriate software allows the data to be retrieved, tabulated, mapped and reformated for use with other software packages. Further, the problems of the dissemination and use of statistical data base is world-wide. Therefore, the application of CD-ROM and PC technologies to the solution of this problem has great significance for developing countries. However, even without these relatively new technologies, because of the availability of mainframe and mini-computer facilities, the statistical data bases of developed countries will still be used to a modest extent. This is not the case with developing countries. The application of CD-ROM and PC technologies is probably the only way for persons to have access to and be able to use effectively their statistical databases for education, planning and for research purposes. In order to achieve this, a close co-operation with the national statistical agencies and firm developing related software and producing CD-ROM is necessary. Conventional data sets may require some restructuring.

The market potential for CD-ROM is enormous. These products can be sold to taluk/tehsil, district, state, and central government agencies, market research and advertising firms, banks, insurance companies, special interest groups such as charities, social planning organisations, large retail establishments, the print and electronic media, political parties, libraries, schools and universities. It is no longer necessary to have expensive mainframe or mini-computer hardware, computer operators, systems analysis and high level language programmers to be able to use statistical data bases. This can now be achieved through the joint application of the relatively inexpensive CD-ROM and PC technologies.

One of the major application area is the censuses. Large data bases for a country like India can be stored on CD-ROM and be retrieved by PCs. India has the tradition of producing censuses regularly and at national, state or union territory and district levels, and there are several volumes at each levels as well. Other censuses, such as livestock or industries can also be recorded on CD-ROM. Space – Time Research, a Melbourne based Company, has done exceptional work in this field. It has placed the entire 1981 and 1986 Australian Census in a single CD-ROM. The data is equivalent to 314 megabytes of information, or about 160,000 A4 pages. The census information can be retrieved instantly and displayed pictorially, which reveals patterns of income and population density and other characteristics such as the age of the population within a municipality and the incidence of religious beliefs (Fig. 12.2). It has developed a PC software package called *Supermap* which together with the CD-ROM is included in CDATA 86. The creation of CDATA 86 involved storing several databases from the 1981 and the 1986 Australian Censuses of Population and Housing on CD-ROM. Digital map boundaries are also stored on the CD-ROM, these data were made available by the concerned survey and land information officer. Now the 1980 U.S. Census data is also available on such disc (Massey, 1987; Stirling, 1988; Coppock, *et al.*, 199; Marx, 1987).

CD-ROM and PC technologies have been used in other countries as well. The National Land Survey of Sweden has produced an experimental CD-ROM disc to see whether it provides a suitable medium on which to distribute its cartographic databases, this disc holds sample maps of the various maps series in Sweden at scales from 1:10,000 to 1:2 million. Another experimental disc has been produced by the National Geophysical Data Center at Boulder (Coppock *et al.*, 1988). The emphasis is on a spatial data. Similar to *Supermap*, the software developed for this purpose can provide not only displays of the data and analysis of them, but also graphs and other graphics. There are also several commercial initiatives to make maps available on such discs, such as DeLorme World Atlas. This disc holds maps at a hierarchy of scales, from the global to the local.

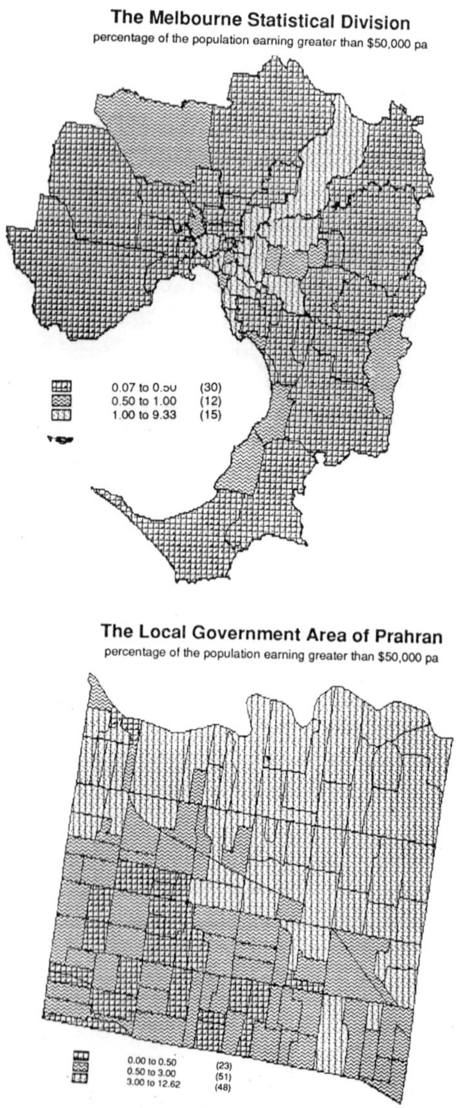

FIG. 12.2: Applications of PC/CD-ROM Technologies

The Indian Case

Considering the merits, demerits and potentialities, it must be assessed how for the combination of PC and CD-ROM technologies is useful for a country like India. Should a developing country go for such a sophisticated system? How far the dissemination of the voluminous census and other socio-economic data has been possible by the present conventional method of publishing reports and hand books? Further, how such data can quickly be processed and used for thematic mapping? These questions will have to be looked into before introduction such technologies.

India is a vast country having 31 states and union territories, or about 470 districts, or about 3,500 sub-district level administrative units. There are about 4 thousand towns and about 570 thousand villages. The population, housing, landuse, agricultural production, livestock, industry related data are collected on the basis of this administrative framework. In the following Table 12.1, an idea about the volume of data can be gathered. Further, there are location specific data which is also sizeable. The examples are of ports, meteorology, industry etc. (NATMO, 1986).

TABLE 12.1: Volume of data

	Information	Level and above	No. of parameters
1.	Landuse	Village	20
2.	Crop use	Village	75-100
3.	Irrigation	Village	75-100
4.	Population and Housing	Village/Towns	80
5.	Agricultural Production	District	75
6.	Fertilizers	District	25
7.	Livestock	Sub-district	125
8.	Agri. Machinery etc.	Sub-district	50
9.	Fishing	Sub-district	20
10.	Ports	Portwise	20-25
11.	Railway	Station	80-90
12.	Mines and Industries	Mine/Industry	100
13.	Health and Vital Statistics	Towns/district	25-30
14.	Forest and Wildlife	Range/Circle	50-60
15.	Agri-Market & Storage	Location	40-50

The case of Indian Census can be cited here as well. The Registrar General of India produces several volumes at national, state & U.T. and district levels. There are specialized volumes as well. The volume of data is to huge that by the time the census reports are published, there is a lapse of about 5 years. A quick dissemination system is most essential in order to use the data effectively (Census of India, 1988).

The other data sets are equally huge and in some cases the frequency of data collection is faster, e.g., the livestock is produced at 5 years interval. Since thematic mapping requires current data, a PC/CD-ROM combination can be useful in producing the maps within reasonable time.

The other use of this combination is to produce CD-ROM containing thematic maps. The large sets of maps are unattractive, and difficult to handle, but the size can not be reduced due to several considerations, such as the thematic contents which is appropriate at a given scale. The example of the *National Atlas of India* can be mentioned here. The whole set is divided into 8 volumes of 300 plates and weights about 40 kg. It has to be packed in a wooden crate for transportation. With the combination of PC/CD-ROM technologies, the plates of this set can be divided into smaller plates so as to fit in single screen. The set of 40 kg can be brought down to a multiple of 18 grams of CD-ROM.

The third possible area of introducing the combination of PC/CD-ROM technologies is the production of satellite remote sensing data. Indian has already launched IRS-IA remote sensing satellite and is planning to send another one soon. The quality of data is comparable with other remote sensing satellites now in operation. Earlier the data was available on CCTs but now it is available on floppies as well, which can be handled by micro computer/PC based system. Such data can now be permanently stored in CD-ROM. There are two problems to achieve this: (a) standardization of data, and (b) development of related software. Standardization will involve rectification, removal of noise and extrapolation of information for the area covered by clouds etc. Most appropriate time of the year has to be worked out so that the maximum number of users find it acceptable. The development of the software relates to the arrangement of data to be store in the CD-ROM so that it can be easily retrieved.

Compact discs have much of the market for recorded music, but it is only in the past two years that the digital technology has been successfully used by the computer industry.

Space-Time Research has placed the entire 1981 Australian census and the 1986 census on to a single CD-ROM, which can be assessed by personal computers. The census data is equivalent to 314 megabytes of information, or about 160,000 A4 pages.

The census information can be retrieved instantly and displayed pictorially, which reveals patterns of income and population density and characteristics such as the age of the population within a municipality and the incidence of religious beliefs.

The information now available on a 12-centimetre CD-ROM, which uses a $1500 CD-ROM reader and a $5000 personal computer, gives the operator access to data previously only available through mainframes.

After an injection of capital by Sydney-based First MIC Ltd, and many months of intensive research, Space-Time Research team were able to produce a software package, known as *Supermap*, which enables the mass data stored on the CD-ROM to be reached and then displayed graphically.

An important feature of the software is that it accesses the information on the CD-ROM at record level, rather than through the traditional file format. This means that the CD can be read more rapidly using a cheap microcomputer than would be possible using most mainframes. Only three companies in the world have developed the software to access a numerical database from CD-ROM in this way. Space-Time Research wants to dominate the alpha-numeric CD-ROM memory market — or any other high-density storage technology that emerges.

Supermap 1.0 has been sold to government authorities, political parties, advertisers, market researchers, libraries, retailers, social planners and churches. Retailers have used the information on *Supermap* to identify income levels and other population features within particular areas to make decisions about new store locations. Advertisers and market researchers have used *Supermap* to plan advertising campaigns, particularly direct marketing.

The census bureau will market and sell the CD-ROM disc containing the census details, known as CDATA 86, and the associated *Supermap* 2.0 software.

Conclusion

The PC technology has already come in a big way in India. Advantage of this situation can be taken. CD-ROM player and production of discs are required. In most of the cases, the data producing organisation have either already gone for the digital recording of data or have such a programme in near future. The data will be available on CCTs in any case. The situation is perhaps ripe to consider the introduction of PC/CD-ROM technologies in thematic cartography in India.

REFERENCES

Census of India, (1988). *Introducing the Census of India Publication*, Registrar General of India Office, New Delhi.

Cooke, Donald F., (1987). Map storage on CD-ROM, *Byte*, July, pp.128-38.

Coppock, J.T.; G. Anderson; T. Banko; J.P. Grelot and P. Nag, (1988). *Report on Thematic Maps of the World*, ICA/IGU/ICSU Working Group (Unpublished).

Dutt, G.K and P. Nag, (1987). *National Atlas and Thematic Mapping Organisation in 2001 A.D.*, NATMO Occasional Monograph 2, Calcutta.

Marx, R.W., (1987). *Report of the ICA Ad-hoc Commission on Population Cartography*, Intenational Cartographic Association, Morelia, Mexico.

Massey, Jack, (1987). *Supermap Uses' Guide and Reference*, Space Time Research Pty Lt., Melbourne.

Nag, P., (1984) (ed.). *Census Mapping Survey*, Concept Publishing Company, New Delhi.

N.A.T.M.O., (1986). *Report on the Digital Mapping System for NATMO*, Department of Science & Technology, Calcutta.

Stirling, Peter, (1988). Read-only software ready for U.S., *Business Review Weekly*, February, pp.58-9.

13

GLOBAL THEMATIC MAPS OF THE NATURAL ENVIRONMENT

J.T. COPPOCK

Between October 1986 and April 1988, a Small working group of the International Geographical Union (IGU), attached to its commission on Geographical Education, investigated the feasibility of producing a set of thematic maps of the world illustrating features of the natural environment and intended to meet educational needs. Its report was submitted in July 1988 and this chapter summarises two elements of its deliberations, the sources available for such a series and the methods whereby it might be produced.

The Remit

The concept of such a series had emerged from discussions at the first symposium on global change of the International Council of Scientific Unions (ICSU), held in Ottawa in September, 1984 (Malone and Roederer, 1985) and one of the precursors of the International Geosphere-Biosphere Programme (IGBP) adopted by ICSU in September, 1986. The aim of the enquiry gradually crystallized as an examination of the feasibility of producing a set of 12-20 authoritative maps illustrating features of the world's terrestrial, oceanographic and atmospheric environments, intended to serve primarily a university-level market but with the needs of developing countries particularly in mind.

In view of the limited time and resources at its disposal, the working group concentrated on two closely-related aspects, the sources available and the technical possibilities for producing the series. That such maps were desirable was implicit in the remit and resources available did not permit any market research to be undertaken, although it was clear that no map publisher had so far identified a profitable commercial market for maps of this kind.

The Sources Available

There were several possible sources for the series. Ideally, maps should be compiled from the original data, as with the climatological maps compiled by C.J. Willmott and C.M. Rowe (1985); but this approach was impracticable for a set of maps covering many features of the natural environment and prepared within a reasonable time scale. In any case, identifying the sources, evaluating them and assessing their mappability would be well beyond the capabilities of this enquiry, (Panel on World Data Centres, 1987). A sub-set of such data, those already in digital form (many of them derived from sensors mounted on satellites), was worthy of further consideration, but identifying and evaluating these sources were also major tasks (see, e.g., Horowitz and King, 1988). There is undoubtedly an immense amount of such data, particularly relating to the atmospheric and oceanographic environments, but no comprehensive record exists and only a small proportion appears to have been mapped, at least in published form. Nor would all imagery be suitable for the production of maps of the whole globe. Much is at too fine or too coarse a resolution (e.g., that from the LANDSAT series of satellites and that used for climatic modelling respectively), and most global coverages, especially of the land surface, are only partial because of cloud cover. At the time of this enquiry, the United Nations Environmental Programme's GRID database contained only seven digital global coverages (Global Environment Monitoring System, 1988) and a preliminary list of environmental coverages prepared by the National Geophysical Data Center at Boulder, Co., contained 36 titles, (including those already in GRID), nine of which were from satellite imagery.

Identifying what imagery had been mapped proved even more difficult. Such maps are not separately catalogued in map libraries, and the agencies collecting the data appeared to have no comprehensive record of what has been mapped from their data. The list which the group identified, totalling 16, is largely a product of chance encounters, mainly in journals and on visits. Such sources have the merit that the data can be readily recompiled to a common scale and projection, and several coverages have been included in the provisional list of 20 maps.

Compiling a list of global data which already exists in printed map from was, in principle, much easier. An incremental approach was adopted, based on searches through the map catalogues in national and specialist libraries (beginning with the Library of Congress), bibliographies and publishers' catalogues to identify all maps at scales between 1:5 and 1:40 million that had been published since 1970, the latter being regarded as the most appropriate scale at which a useful map of the whole globe could be published on a single sheet. In retrospect, that lower limit was one the group came to regret because there are many smaller maps at scales up to 1:150 million, notably in specialist atlases, which might be suitable for representation on other forms of output. In all, over 600 maps were identified by these searches of which 150 were regarded as worthy of further consideration because of their content.

The thematic coverage of these maps was, however, very uneven, with nearly half the maps portraying some feature of the earth's geology or geophysics. There were remarkably few showing soil and vegetation and a surprisingly small number portraying the oceans and the atmosphere (though many more would have been identified in the atlases mentioned earlier). There were several other weaknesses in such an approach. In all major collections there is a backlog of varying seriousness of items awaiting cataloguing, so that some of the more recent maps may have been missed. There is, moreover, no international equivalent of national copyright libraries, entitled to receive a copy of all publications, and holdings of global maps consisted of donations and of selected purchases of known maps. Nonetheless, there is evidence to indicate that the list contains a substantial proportion of all published maps, although it will need progressive updating as new maps and new editions appear. Whether there are many unpublished maps, is, of course, unknowable, although they certainly exist, particularly in relation to features of the atmosphere.

Once possible sources had been identified, it became necessary to select 20 coverages, as required by the remit, and the following criteria were used to aid selection: the scientific integrity of the map; its global consistency; its up-to-dateness; its scale and complexity; the environmental significance of the topic; its educational value and interest; and whether it was free of copyright. An attempt was also made to secure a balance between different features of the global environment. The 20 were not, however, those that would have been chosen in an ideal world, which, would have included many portraying such aspects as rates of change and variability.

Possible Methods

Four possible approaches to the production of a set of global thematic maps were considered, one leading to printed maps, the others to non-traditional forms of output.

The simplest and cheapest solution would be the preparation of colour microfiche of each of the selected maps (see, e.g., Color Micro Imaging Corporation). The quality of such microfiche has improved greatly, enabling maps as large as 155 cm by 125 cm to be photographed without loss of resolution; they are cheap (at under US $50 for a master and US $1-2 or less per copy in one commercial system), as is the equipment for viewing them replacements and extra copies can be produced quickly; and microfiche are cheap to store and distribute. Most importantly, no recompilation is required. On the other hand, the maps being photographed would be at different scales and projections; they could be viewed only at a small scale in a viewer or at a large scale when projected onto a screen (neither of which would be wholly satisfactory for the intended market); and they have not attracted much attention from commercial publishers. On balance, this method did not seem suitable for this project, although microfiche could be considered as a supplementary medium to facilitate storage and consultation once a set of maps on a common scale and projection has been prepared.

Another photographic approach, recording on videodish, likewise avoids the necessity of recompiling maps and has attracted some commercial interest, as in the FULCRUM system provided to the U.S. Department of Defence as a facility for rapid consultation of maps in the field (Interactive Television Company, 1987) and the British Broadcasting Corporation's Domesday project, where videodisks are used for reference and for educational purposes (Openshaw and Mounsey, 1987). A videodish holds up to 55,000 images on a single side, although one map may require a number of images; both the systems mentioned above hold maps in a hierarchy of scales. Once the necessary maps have been assembled, disks can be made quickly; maps can be accessed rapidly; limited software exists to permit some overlay of other data and disks are robust and easily transported.

On the other hand, except for small maps, only part of the global map can be seen at any one time on the colour monitor, and relatively costly equipment is required by users, viz., a videodisk player, a microcomputer and a colour monitor together with appropriate software, totalling US $17,000 in one commercial system (with $25 for each disk once a master had been made). In view of the remit to have regard to the needs of developing countries and the limited viewing area, this approach was also considered inappropriate for this project.

The third approach, which was likewise rejected on grounds of cost to users, is the recording of maps on compact disc read only memory (CD-ROM). This is a rapidly-developing technology which has already been used in a number of commercial cartographic applications, such as SUPERMAP (Massey, 1987) and is under consideration by several mapping agencies and scientific organizations (e.g., B. Rystedt, 1987). Its main advantages are that, because the data are held in digital form, they can be manipulated in a microcomputer in various ways, such as changes in scale and projection or overlays with other data; a single disc can hold as many as 550 megabytes of data; discs are robust and easily transported; and the technology will benefit from its use in the much larger market for sound recording. Its principal disadvantages are that analogue maps must first be digitized; the display area is restricted by the size of the colour monitor (although complete maps can be output to a plotter); and the system is relatively costly to the user, requiring a CD-ROM reader, microcomputer and monitor.

Such equipment will cost US $10,000 for a working system and, after mastering, each disc can be produced for c. US $10. At present such discs cannot be updated, but it is thought likely that this limitation will soon be overcome. Although this approach was rejected for this project, the group regarded it as the most promising of the non-conventional approaches and one that should be developed in parallel to the production of printed maps.

There was an implicit assumption in much of the preliminary discussion about this project that output would be in the form of a printed maps on a common scale and projection. The main advantage of this approach would be the use of well-tried technology to produce a familiar product, use of which would not require access to expensive equipment. Its main disadvantages are that the product requires recompilation to a common base; maps can be updated only by making new editions; they can only be inspected visually; and maps are costly to store and distribute. The estimated costs of production varied widely with the approach and complexity of the source map, but most were in the range of US $20-30,000, or approximately $10 per sheet.

Three different approaches are possible to the production of a printed map; (a) a wholly manual approach, by which maps are recompiled onto a base map, from which colour separations are made for printing; (b) a wholly digital approach, in which the source map is first digitized, then transformed to the required scale and projection, edited on line and then printed, a process not requiring colour separation; (c) a hybrid approach, in which the source map is recompiled onto a base map. The recompilation is scan digitized and the map printed, again without the need for colour separation. The digital approach is the most flexible, but is also the most expensive and can be justified only where different series are to be produced from the same data. Although the manual approach is well tried, substantial savings could be made on printing costs by using the hybrid approach, and it was this that the group recommended after intensive consultation with map makers.

The Market for a Series

The great majority of thematic maps of the kind considered in this chapter have been produced by public agencies, both international (e.g., UNESCO) and national (e.g., the U.S. Geological Survey). Most maps have been published for a variety of educational, scientific and non-commercial purposes and provide only limited guidance on the market that might exist for a set of thematic maps. The global maps of geological topics produced by the National Geophysical Data Center have generally had a print run of 1-2,000 copies; they were formerly distributed without charge on request, but a small charge is now made.

Numbers distributed have varied widely; maps are still available for some themes and others have been reprinted in response to demand. Such maps were publicized through leaflets, mainly within the United States, but were not marketed in any commercial sense. Other organizations and enquiries gave similar indications of widely-differing demands for individual sheets. As a conservative and very subjective estimate, the group thought that 1-2,000 sets worldwide might be distributed, but a great deal would depend on the price charged (if any) and the efforts devoted to marketing. To give a better basis for estimating costs and markets, the group recommended that a pilot study be undertaken with one or two maps. It also recommended the development of user-friendly software for the CD-ROM-based maps.

Identification of the likely size of the market is further complicated by the existence of other initiatives and proposals, some of which came to light in the course of enquiries. The Soviet Academy of Sciences has in hand an atlas of the natural environment which will contain 154 global maps at scales of 1-60, 1-80 and 1-150 million and 197 continental maps at scales between 1: and 1:60 million. The Smithsonian Institute is planning a global atlas of satellite imagery; and the GRID database, now established on a permanent basis, intends to increase its holdings.

The Commission of the International Cartographic Association on the World Digital Database for Environmental Science has secured funding for the preparation of a digital map framework based on the 1:1 million Operational Navigation Charts, and the Defense Mapping Agency in the U.S. plans a similar initiative. The National Geophysical Data Center at Boulder has proposed a distribution of existing global environmental data sets on CD-ROM discs and the publication of a hard-copy atlas derived from the same source (Kineman *et al.*, 1986), the National Aeronautics and Space Administration's (NASA), Earth Systems Science Committee (1988) has recommended an extensive programme of data collection and the finalization of the I.G.B.P. programme in September 1988 will no doubt lead to other initiatives to improve the coverage of global data. In anticipation of these developments, the I.G.U. has established a Global Data Planning Project to help ensure that these data are used efficiently (Mounsey, 1988). No doubt these and other initiatives will lead to a variety of products over different time scales and at different levels of resolution. What seems important, particularly for the scientific community with its limited resources, that there should be the maximum possible co-operation between these various initiatives.

REFERENCES

Earth System Science Committee, (1988). *Earth System Science, a Closer View*, National Aeronautic and Space Administration, Washington.

Global Environment Monitoring System, (1988). *GRID Pilot Phase 1985-87. Final Report*, United Nations Environmental Programme, Nairobi.

Horowitz, R. and King, J.H., (1988). *NSSDC Data Holding*, 88-01, Goddard Space Flight Center, Greenbelt, Md.

Interactive Television Company, (1987). *FULCRUM Electronic Map Display, Videodisc Map Technology*, Arlington, Va.

Kineman, J.; Hastings, D. and Colby, J., (1986). *Developing global databases for the environmental sciences*, in Proceedings, 20th International Symposium on Remote Sensing of Environment, Nairobi.

Malone, T. and Roederer, T.F., (1985). *Global Change*, Cambridge University Press, Cambridge.

Massey, J., (1987). *SUPERMAP Users' Guide and Reference*, Space-Time Research Pty Ltd., Melbourne.

Mounsey, H. (ed.), (1988). *Building Databases for Global Science*, Taylor & Francis, Basingstoke.

Openshaw, S. and Mounsey, H., (1987). Geographical information systems and the BBC's Domesday interactive videodisk, *International Journal of Geographical Information Systems*, vol. 1, pp. 173-9.

Panel on World Data Centres, (1987). *World Data Centres*, Part 1, International Council of Scientific Unions, Boulder, Co.

Rystedt, B., (1988). Compact discs for the distribution of maps and other geographical information, *Proceedings, 13th International Cartographic Congress,* 4, Morelia, pp. 479-84.

Willmott, C. and Rowe, C.M., (1985). Climatology of the terrestrial seasonal water cycle, *Journal of Climatology*, vol. 5, pp. 589-606.

14

FIFTH GENERATION COMPUTERS AND BEYOND

R. THIAGARAJAN

Computers, like human beings, are being classified according to generations. Starting in the late 1940s as number-crunching machines, their (computer) designers are increasingly being called upon to build machines that can recognise symbols, hand writing, spoken words on the one hand and perform a host of other chores performed by human beings on the other.

Computers are also being used for different activities related to thematic cartography. Digitization, development of data bases, plotting, symbolization, editing, manipulations, digital image processing and digital mapping can now be done on computer-based systems of varying size and capacity.

It would be worth mentioning that thematic cartography will not be left untouched from the progress made in computer hardware, software and applications of various hue and colour. Large memories, parallel processing, optical discs and the like will make thematic mapping technology easy but more specialized. Hence, it is in the fitness of things that we look at the changes likely to occur in the field of computer technology as well.

While we, as humans, have been using the earliest form of computers (the ten fingers) given to man by God since times immemorial and, progress in computing devices has been slow for ages, we seem to have progressed in the use of electronic devices for number-crunching operations during the second-half of the 20th century by leaps and bounds. Computers, like human beings, are improving, of late, in their capability, storage capacity and indeed in truly assisting man in his mental faculties.

The co-ordinated research work undertaken by the Japanese (since

1979) in the development of the Fifth generation is indeed remarkable and needs to be emulated by the Indian computer community. There are a number of reasons why Indians (and not necessarily others) should have a greater urge to contribute to the development of the next generation of systems (not necessarily computers in the strict sense of the term). Firstly, the Indian seers and thinkers, since the dawn of civilization, as it were, have engaged themselves in answering questions such as – "What is thinking?" "How does one store experience in the brain?", "Why does one forget?", and so on. Secondly, if anything, Sanskrit alone (of all the world languages) seems to get anywhere near the human-machine interface. Thirdly, whether Sanskrit is versatile enough or not, the grammar evolved by Panini for Sanskrit needs an in-depth study before any further work on languages for the fifth and other future generations is undertaken.

This chapter attempts to give an overview of the following:-

(a) Generations of computers and their characteristics,
(b) Expert tasks and systems,
(c) Chronological sequence of events in the growth of the fifth generation,
(d) Basic configuration of the fifth generation and its description.
(e) Current status of the fifth generation,
(f) A comparison between Von Neumann systems and the fifth generation, and
(g) Some comparisons between the human brain and the computer or an "Information Engine".

The Computer Generations

Starting with a project on an electronic computing instrument in 1946 at the Moore School of Electrical Engineering, Pennsylvania, electronic computers have made a mark on human kind over the years. The first generation computers can be said to have been born with the publication of a paper entitled "Preliminary discussion of the logical design of an electronic computing instrument" by John Von Neumann and others which highlighted the salient features of the computer project. Though parallel work was going on in the U.K. under the direction of Alan M. Turing at the same time, almost all computers developed till date (till the Fourth generation at least) are called Von

Neumann machines.

The various generations of computers till date could be compared under the following headings (as shown in Table 14.1):

(a) Year of introduction,
(b) Type of components for storage,
(c) Number of users,
(d) Type of languages used,
(e) Nature of inputs,
(f) Nature of processing,
(g) Cooling arrangements, and
(h) Space requirements,

In addition to the above, the fact that the cost of systems for the same computing power has been falling enormously is also a factor to be considered while comparing different generations.

We are now in the "Fourth Generation" era. So far, one has been according the status of a new generation whenever there was improvement in circuit components; design of new languages or when there was any change in telecommunications technology. Underlying all this was the philosophy that all computing machines were meant to compute (in other words crunch numbers faster and faster).

The Fifth Generation, which is going to be described later, is expected to be a radical departure from the earlier generations in that the V generation would handle knowledge instead of data as in the earlier generations.

The Sixth Generation, which is yet to reach even the "Concept-forming stage", ought to, if anything, introduce biological storage devices with natural language interaction using voice and sight inputs. Even though voice inputs and pattern recognition have been given prominence in the Fifth Generation itself by the Japanese, one wonders whether these goals will be reached during the decade of the eighties itself.

It would be appropriate now to digress a little and see how the human brain compares with the computer.

Computers and the Human Brain

Attempts have been made to compare the computer with the human brain from time to time. A.M. Turing, Norbert Wiener and others have a significant thrust to the concept of the computer taking on more and more human thinking functions. Thus, the field of Artificial Intelligence came into existence. In fact Turing's article "computing machinery and intelligence" published in *Mind* in 1950 could be said to have launched Artificial Intelligence (AI) as a separate branch of study.

Let us now look at some of the features that distinguish a computer from the human brain as of date. A brief comparative table between the two is given in Table 14.2. In the main, there does not appear to be any solution in sight in the near future to make computers that would take over the major capabilities of a human being. Yes, motor functions such as limb movement, recognition of alphabets, and to some extent, speech recognition and output seem to have reached a fair degree of sophistication where machines are concerned. Such functions are usually clubbed under the "Perceptron approach". The thought process, the whole host of activities that are performed in creative work etc. seem to be beyond the pale of machines as of now. A number of researches the world over have been engaging themselves in the design and development of A1 systems.

Artificial Intelligence and Expert System

What is AI?

The term Artificial Intelligence (A1), was coined by Dr. John MC Carthy, Assistant Professor of Mathematics in Dartmough College in 1965.

A number of definitions have been given for AI. These are –

(a) By Marvin Minsky of MIT - "At is the science of making machines do things that would require intelligence, if done by men".
(b) By Feigenbaum (Inventor of Dendral) and Mc Corduck – "If we can imagine an artefact that corduck – "If we can imagine an artefact that can collect, assemble, choose among, understand, perceive and know, then we have AI".
(c) By Aaron Sloman – "Theoretical analysis of possible effective explanations of intelligent behaviour; explaining human abilities and constructions of intelligent artefacts.".

TABLE 14.1: Comparison of Generations

Sl.No.	Facet	Generation I	II	III	IV	V	VI
1.	Year of introduction	1946	Early 50s	Early 60s	Early 70s	1982	1990 ?
2.	Storage circuit Components	Thermionic valves	Transistors	ICs	LSIs	VLSI	Biological ?
3.	Number of users	1	1	Multiple user community	Remote users	Same as IV generation	Not known
4.	Languages used	Machine Language (low-level)	Symbolic Language FORTRAN (High-level Language)	COBOL, ALGOL, PL/a etc.	PASCAL, Query Word Processing	Kernel Language, Query Language etc.	Natural Language of the user ?
5.	Nature of inputs	Torn tapes	Punched cards, MT, Disks etc.	Cards, MT Disks.	VDUs, Floppies, Optical Disks, COMs.	Optical Disks, Optical records, waves ? Speech input.	Speech, Signs, Thought
6.	Nature of processing	Serial	Serial	Serial	Serial/ Parallel/ Array	Parallel/ Array	Associative ?
7.	Cooling arrangements	Extensive	Moderate	Less	Room A/C adequate	Room A/Cs	In-built liquid coolers ?
8.	Space requirements	Large	Medium	Small	Normal office room. Table top PCs	Office room Table top Palm-top	Hand-held ?

A1, as shown in Figure 14.1, encompasses a whole host of activities, principal among them being development of expert systems. Let us now see what Expert Systems are by looking at a scenario.

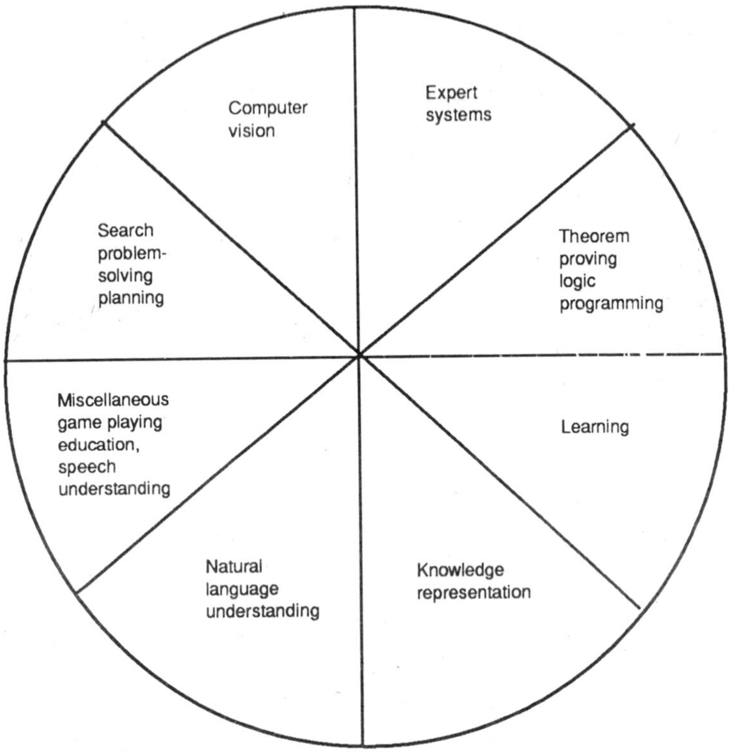

FIG. 14.1: The "Artificial Intelligence Pie"

Imagine a situation when the collector of a particular district, (say, Midnapore), wishes to find out the quantity of the rice, atta, sugar, tea leaves etc. required for flood victims. The Artificial Intelligence package (let us call it, "swa – ludhi", for short) should be able to not only give him all this information superimposed on a topographical map of the district but also order the necessary items from governmental go-downs to be sent to Midnapore. The "Swa – ludhi" package must be connected to the state-wide network and thus be able to order helicopters to take off from helipads; boats and crew to be sought from the nearest army camp/police camp; doctors to be flown to flooded areas; medicines to be airdropped etc.

Thus, there is a strong case for developing an Expert System (say, "Swa-Ludhi") to be able to handle management of disasters such as floods, famine, earthquakes, landslides, etc. Here is where the expertise of a "Thematic Mappist" comes to the fore.

Let us now see what expert systems are.

TABLE 14.2: **Some comparisons between the computer and the human brain**

Sl. No.	Computer		Human brain
1.	Non-living device	—	A living organism.
2.	Dependent on humans for its programme	—	Self-willed.
3.	Discrete devices	—	Analogue devices operating on a continuous mode.
4.	Semi-conductor, VLSI storage	—	Electro-chemical form of storage or is it electro-magnetic chemical (EMC) storagel ?
5.	Portions lost are lost for ever	—	Possesses in-built capacity to reactivate lost portions of memory.
6.	Memory size no problem as extra memory could be added on	—	Limited memory size (expected to be 10 Billion neurons)
7.	Very fast in operations (latest is of the order of 100 Megaflops or 100 Million Instructions Per Second (MIPS)	—	Speed of operations rather slow-of the order of a few milli-seconds.
8.	Speed of transmission = Speed of light	—	Speed of transmission of the order of 100 to 120 metres per second.
9.	Only retention and recall have achieved a certain degree of sophistication	—	One of the finest and most compli-cated devices with respect to intelligence (observe, Retain, Recall, and Generate ideas).

Expert Systems

Expert systems basically provide a repertoire to a computer with regard to the mechanics adopted by an expert in the fields of medicine, engineering design etc. A list of expert systems, along with their objective is given in Table 14.3. Such "expert systems" make use of stored "knowledge-base" for solving complex problems in specialized fields, and include facilities such as automatic learning and inference.

Two of the expert systems developed so far need more attention. These are :

TABLE 14.3: Representative Expert Tasks and Systems

Some of the expert systems developed for accomplishing various activities.

(1) *Interpretation*

Purpose: To find a consistent and correct interpretation of data (e.g.) measurement data.

Systems

(a)	DENDRAL	: Interprets mass-spectrometer data.
(b)	DIPMETER	: Analyses oil-well log.
(c)	PROSPECTOR	: Evaluates sites for mineral deposits.
(d)	WAVES	: Advises on seismic data.

(2) *Diagnosis*

Purpose: To locate faults in a mechanical or human system.

Systems:

(a)	MYCIN	: Meningitis and blood infections.	Medical diagnostic systems.
(b)	PUFF	: Lung disorders.	
(c)	CADUCEUS	: Internal Medicine.	
(d)	INTERNIST	: Internal Medicine	
(e)	DART	: Computer System faults diagnosis	

(3) *Monitoring*

Purpose: To critically follow behaviour to warn about deviations from norms and suggest corrective measures (diagnosis on-line and real-time).

Systems

(a) VM: Monitoring a mechanical breathing device.

(4) *Prediction*

Purpose: Prediction of future behaviour of a system based on past history and present status.

(5) *Design*

Purpose: Design of a system to realize a specified behaviour.

Systems

(a)	KBVLSI	: VLSI design assistance.
(b)	SACON	: Structural Engineering Analysis.
(c)	PROGRAMMER'S ASSISTANCE	: Assistance in software design and debugging.

Source: Prof. R. Narasimhan, NCSDCT, TIFR, BOMBAY - 400 005.

(a) *Dendral*

This was designed in 1965 for use by organic chemists to infer the molecular structure of complex organic compounds from their chemical formulae and mass spectrograms (mass spectrograms are mainly bar plots of fragment masses against the relative frequency of fragments at each mass). To do this, DENDRAL makes use of rules which relate physical features of the spectrogram to the need for a particular substructure to be present or absent from the unknown chemical structure. This programme is claimed to rival expert human performance for a number of molecular families.

(b) *Mycin*

This system diagnoses the causes of some bacterial infections (especially in the blood) and recommends appropriate drug treatment on the basis of an interactive dialogue with a physician about a particular case.
To do this, MYCIN makes use of a "Knowledge Base" of over 400 rules of the form –
IF (condition) THEN (implication). Each of these has an associated "degree of certainty" indicating a subject expert's level of confidence in the rule.

The use of rules of this kind, as a means of representing complex knowledge, is typical of expert systems, or at least of the kind so far developed. Not only do such rules enable inexact (or heuristic) information to be included, but they are also presented and stored in memory in an explicit form, which is simply not possible with a conventional programme in, for example, FORTRAN or COBOL.

More about Expert Systems

In addition to the ability to solve problems, expert systems are generally able to explain, and even discuss at length, their own chains of reasoning; for example, what rules acting on what clinical data led to a particular diagnosis? Again, the explicit encoding of knowledge in rules makes an English language explanation possible in a way that is virtually impossible for a programme in a conventional computer language.

Development of PROLOG

A major tool for assisting computer science workers in the field of A1 which has been developed in the recent past is a language called PROLOG. Prolog is essentially *a rule-based programming language*, in which a rule (such as the earlier one quoted the MYCIN) can be directly encoded on a single statement. The method used for combining rules is rather restrictive, however, and the language has a number of defects for more complex applications though it is still easy to use, compared to other computer languages.

Von Neumann Machines and Fifth Generation Computers

The Fifth Generation framework, according to Kazuhira Fuchi, the Director of ICOT Research Centre. Tokyo, is intended to reconstruct the body of computer hardware and software on the basis of a logic system called Predicate Logic. In other words, it considers the next generation computer as a *Predicate Logic Machine*. One could call it an inference engine as the basic operation of predicate logic is indeed *inference*.

Current computers, on the other hand, have a language called a machine language. The machine language defines the architecture of a computer. Machine language is also used to build the basic software.

The characteristics of the Von Neumann type machines are intensively represented in the machine language.

Kernel Language

In the Japanese conceptual framework, the machine language equivalent is called the *Kernel language*. A predicate logic-based language, to be used as the Kernel language or machine language, is expected to be formulated. Such a new machine language would necessitate development of new hardware and software.

The FGCS

The hardware in the FGCS will be basically designed for parallel operation or associative search, since its basic function will be inference. Present Von Neumann type computers are mainly built for sequential operation and address search. Thus, the parallel, associative inference machine will be a kind of non-Von Neumann computer (Fig. 14.2).

The software in the FGCS is expected to be built by combining the basic inference functions provided by the hardware. The Kernel Language (KLO, KL1 etc) falls under the category of very high-level language in the current technological framework. The Kernel language is not a user language. Higher-level languages than the Kernel language are expected to be formulated one after another, as user-oriented languages. The present 'knowledge-representation languages' could be positioned at that level, though the ultimate form of higher-level user languages will be natural languages.

Some of the major differences between Von Neumann type computers and FGCS are listed in the Table 14.4. Apart from the factors highlighted so far, one of the important differences lies in the rating of the systems. While Von-Neumann type systems were generally classified by the number of instructions per second, say, millions of instructions per second, also called (MIPS), FGCS will be characterised by Logical Inferences per Second (LIPS).

Beyond the Fifth Generation — Some General Considerations

The Fifth Generation project is just a beginning towards making a machine that may be able to simulate human faculties. The Fifth Generation is said to be a radical departure from the earlier generations, in that the specifications of the system are expected to be drawn up first and then the hardware is expected to be developed, not restricted by the existing hardware.

Another claim that is made is that the Fifth Generation is not necessarily an extension of the current exercises in Artificial Intelligence and Expert Systems though the project is said to be closely related to these fields. It is said that the research is moving towards an integration of software engineering research, database research and artificial intelligence research which have been conducted by independent groups so far (Table 14.5).

There are basically two aspects on which the research into the Sixth Generation ought to be progressing. These relate to :

(a) a recognition that the hardware devices need not be hard electronic circuits such as VLSI etc. and

(b) the fact that if a system is to be developed depending upon identified requirements, then even the fundamental code for numbers needs a re-look.

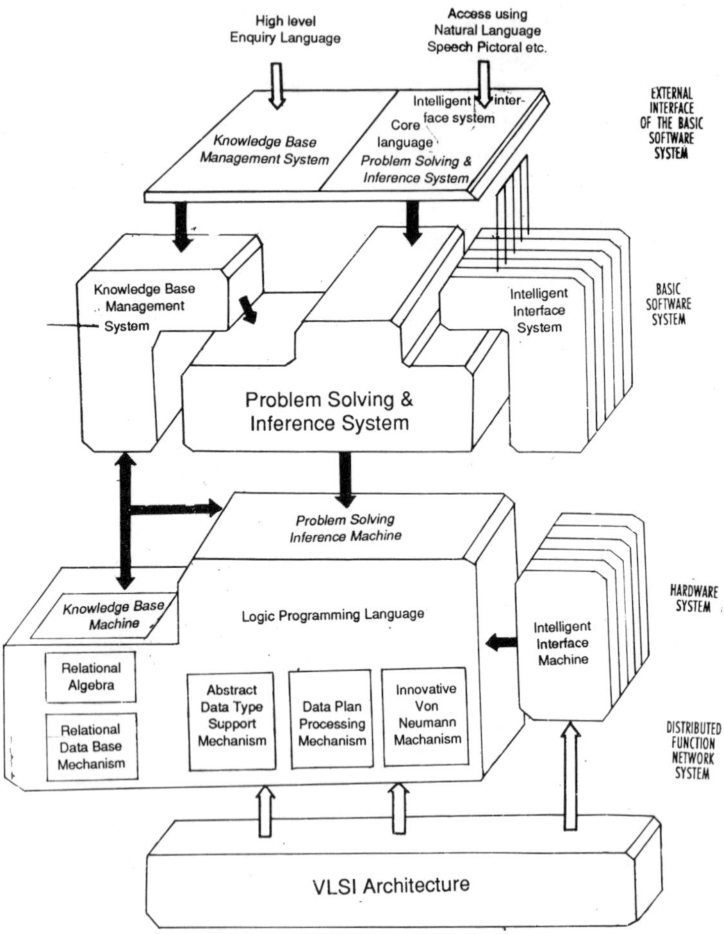

FIG 14.2: Basic Configuration Image of the Fifth Generation Systems

TABLE 14.4: The Von Neumann Connection

The Von Neumann Machine	The Fifth Generation systems
1. A sequential machine (Sequential Operation and Address Search)	Based on Parallel architecture, and Associative Search and its basic function will be Inference.
2. Works on Machine Language which defines the architecture of the machine. Software is basically built on Machine Language.	Works on a high level Language called Kernel Language 0 (KLO) — A non-procedural Language.
3. Has mainly been processing data	Attempts to process knowledge.
4. Rated in Instructions Per Second (MIPS)	Rated in Logical Inference Per Seconds (LIPS).
5. Basic block diagram as given below:	Basic block diagram as given below.
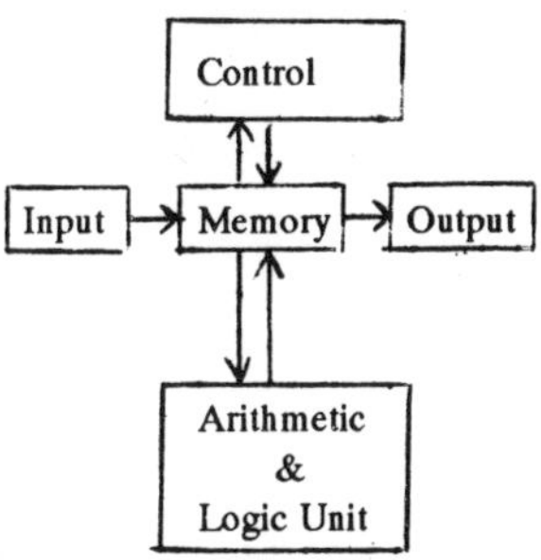	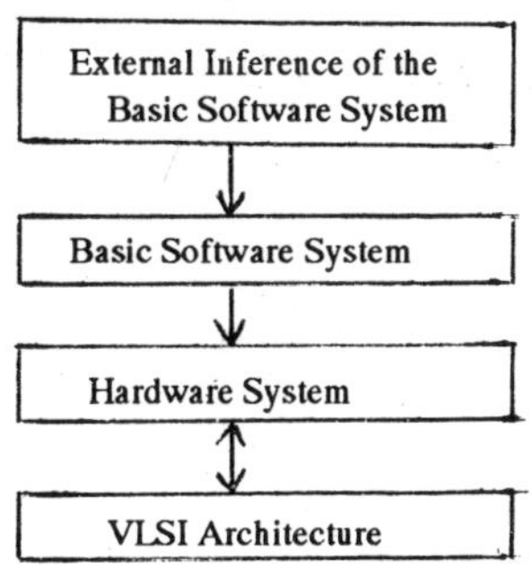

Change from Hard Electronic Circuits

There appears to be a mistaken notion that the neuronal circuits in the human brain are inflexible hardwired circuits (at least the Long Term Memory is considered to be so). The second myth is that the storage and retrieval operations in the human brain obey certain rigid rules and procedures. These two basic assumptions seem to run right through the various generations in the design of computers, and more importantly, Artificial Intelligence Systems (including expert systems) ending with the Fifth Generation Systems.

TABLE 14.5: **Fifth Generation R & D themes**

Sl.No.	System	Theme	Sub-system
1.	Basic application System	1-1	Machine translation system.
		1-2	Question-answering system.
		1-3	Applied Speech Understanding System.
		1-4	Applied Picture and image understanding System.
		1-5	Applied Problem solving system.
2.	Basic software System	2-1	Knowledge base management system.
		2-2	Problem solving and inference system.
		2-3	Intelligent interface system.
3.	New advanced architecture	3-1	Logic programming machine.
		3-2	Functional machine.
		3-3	Relational algebra machine.
		3-4	Abstract data type support machine.
		3-5	Data flow machine.
		3-6	Innovative Von Neumann machine.
4.	Distributed function architecture	4-1	Distributed function architecture.
		4-2	Network architecture.
		4-3	Data-base machine.
		4-4	High-speed numerical computation machine.
		4-5	High-level man-machine communication system.
5.	VLSI technology	5-1	VLSI architecture.
		5-2	Intelligent VLSI CAD system.
6.	Systematization technology	6-1	Intelligent programming system.
		6-2	Knowledge base design system.
		6-3	Systematization technology for computer architecture.
		6-4	Data base and distributed data system.
7.	Development Supporting Technology	7-1	Development support system.

If we wish to make a system somewhat akin to the human brain, the following principles may have to be adopted :

(a) The storage device (neuron) in the human brain apparently acts more like an oscillator on some occasions, an amplifier at others and a filter at yet others. It is possible that experiences are not only stored in the neurons (by its being in a fired state of a quiescent state) but also in the form of waves.

The wave form of storage enables us to explain –

(i) flexibility encountered in storage of experiences,
(ii) forgetfulness in human beings – if the concerned circuits are not tuned to a particular frequency, information stored in that frequency cannot be retrieved, and
(iii) ability to recall an event (however remote in time and space it might be) – which should be possible if the relevant network of neurons were to vibrate at the frequency required to generated or re-create the concerned event.

(b) The code used in the electronic devices so far is a binary code. The question now is whether this code is really useful in a "thinking" environment.

Therefore, one would expect the storage devices of the next generation (say the sixth) to be resilient, handle wave form of energy contained in a piece of information, and yet be orderly and organized according to a scheme.

Parallel or Network Architecture ?

Yet another dimension is added when we get wedded to the concept of parallel architecture. Parallel architecture, on the same lines as sequential access, cannot be the design amenable to associative type of storage of experiences in the human brain. What we need is a network or associative architecture. This architecture will be different from the conventional wisdom on parallel architecture.

What is now needed is a mechanism by which speech input and symbol input could be related to an event through a "network" or an "associative circuit". It is time that designers of the new generation gave some thought to this. Otherwise, all we would achieve is a machine which translates (at least tries to transliterate) from one human language to another – say Tamil to English, Japanese to English or vice versa. Of what use is the storage of a bi-lingual dictionary in the computer devices is a moot question mainly because even humans would not be able to translate from one language to another (one of them unknown) using a book of rules and a dictionary alone.

In the main, what we seem to be moving towards is a *Biological* form of storage with all its seemingly (because of our ignorance) disorderly mechanisms.

A re-look at the number code

Since we wish to move away from the conventional machines and the philosophies underlying them, we ought to be looking at the very number code we have been adopting so far! It seems quite logical that we as humans adopted the decimal notation because God gave us ten fingers. If god had given us eight fingers instead of ten, perhaps we would have adopted the octal system!

When electrical and electronic devices started entering the arena of information storage and retrieval, we adopted the binary system, as such devices were bi-stable in nature.

But when we wish to design a system which could accept inputs like voice, imagery, smell, touch, taste, etc. we may have to look for devices which can be stable in any one of these states (or more). The main question is – how many states should there be ? We may have five states for five sensory inputs with one state for associative storage and another for creative or thought process. We need each one of the storage devices to possess the ability to adopt any one of the seven states because:

(a) it will be easier to standardise on one type for use in all functions, and
(b) it will be possible to switch over from one segment of the memory to another in case of damage.

Therefore, it becomes necessary to use a code with base 7 (say 'sapta', standing for seven in Sanskrit) instead of base 2 of the binary. We may also have to develop devices which possess 7 stable states.

It may not be feasible to make devices with 7 possible states. What should be possible, however, is to make devices which can generate any one of the 7 possible frequencies. Once we have such a device, we should be able to get combination of 8659 frequencies (fundamental, single frequencies; various combinations of two frequencies, three frequencies etc. upto different combinations of 7 frequencies).

Such a methodology enables us to meander our way through the network of storage devices.

Conclusion

This chapter has attempted to describe the various generations, the genesis and current status of the Japanese Fifth generation project and the possible direction in which one may have to proceed for even trying to simulate some of the human faculties. An attempt has also been made to look at "wave form of storage" as also "network access" rather than "parallel access". A new numeric code on which the physical devices would be based using 7 as the base (and called "Sapta Code"), may become necessary. Whether storage elements would be inflexible and rigid electronic circuits or flexible and resilient biological elements would have to be studied in more detail.

In any exercise of the kind engaged by us in this study, one expects hard experimental evidence in support of one's hypotheses. But, the author is of the firm belief that theoretical models must precede experimentation, as otherwise one might try to rely upon one's luck, waiting for a breakthrough. No claim is being made here that experimentation has been tried out. What is, however, worth observing is the fact that machine thinking (if such a term is permitted) cannot possibly be structured on rigid cules! Can we say with confidence that a certain proportion of "nature" (the family into which one is born), with an appropriate proportion of "nurture" (the environment in which one is brought up) would provide the necessary climate for creative work of outstanding calibre even warranting the award of a Nobel Prize?

15

SCENARIO OF THEMATIC CARTOGRAPHY IN 21ST CENTURY IN INDIA

PRITHVISH NAG and R.K. BANERJEE

The challenges of the 21st century for betterment of socio- economic conditions and quality of life would be enormous. Such challenges are more relevant in developing countries like India where desperate attempts are being made at every level, from individual to national, for a better future. But, the problems in the third world countries are already colossal. It certainly requires a higher level of management of resource and environment in order to face the issues of today and challenges of tomorrow. Population in these countries will continue to increase while the environment relationship will be very difficult task for the national and regional planners and resource managers.

The higher level of management can be achieved by interpreting the latest/real time information in correct perspective. Preparation of thematic maps based on such data sets will help in this regard. The issues in thematic cartography are likely to change. New themes based on ever-increasing refined data will have to be introduced. Neither data nor better technology can be ignored by this branch of cartography.

On the other hand, the cartographers in the developing countries will have to get out fast from so-called 'surveying mentality' (Taylor, 1985). Each developing country has its quota of problems in promoting thematic cartography either for planning or for census operations (Nag, 1984a). In some places reliable data is accessible while in other countries trained personnel data are not available. Technology and assistance from foreign countries are not always useful, conversely new set of problems are created (Taylor, 1986). India is not an exception from other developing countries, however it has a long tradition of preparing thematic maps. Indian experience is not necessarily a prescription for the developing countries. Each country has its own policies and practices. The case study of India is likely to indicate the expected possibilities, options and alternatives it may have in the next century and how it is going to meet the challenges.

Methodology

Based on Harva method of formulating long-range future oriented views, S.C. Seth has developed a modified methodology for scenario development which is known as "Seth-Harva Method". This method has been discussed in over thirty national level workshops and has been found suitable methodology for scenario building. The key elements in the Harva method was the envisionment of the future society which has been elaborated in 5 basic steps in the modified methodology. Former method pertains to the futures of organisations and institutions as well, however, some changes have been made in order to meet the specific requirements of this study. According to Seth, there is a scope of having another step by drawing up action plan for current day decision after having visualised a preferred future (Seth 1975 & 1985). The changed scheme is as follows:

Step 1 : A survey of the present state of art in thematic cartography
Step 2 : Trend scenario for the year 2001 A.D.
Step 3 : Future of the Society in the beginning of 21st century.
Step 4 : Preferred future for thematic cartography.
Step 5 : Preferred future scenario for national thematic cartography institutions.
Step 6 : Action plan.

The study can be divided into above six parts. Preferred scenario, technological development, options and alternatives, jump in technology, action plan and the like have been discussed within the broad framework of this methodology.

A Survey of the Present State of Art in Thematic Cartography

Before the publication of the first *National Atlas of India* in the national language Hindi in 1957, the state of art in thematic cartography was nebulous. However, there were piecemeal attempts made the some individuals. Initiatives taken by Professor S.P.

Chatterjee in the form of *Bengal in Maps* and other landuse maps of the Indian side of Bengal are worth mentioning. In a way these initiatives led to the beginning of thematic cartography in the country (Dutt, Nag, 1985).

Following the first atlas, the then National Atlas Organisation took a greater responsibility of producing and enlarge version of the *National Atlas of India* consisting of 300 plates. This was a landmark in the history of the thematic cartography in the country. From time to time thematic atlases or national atlas in parts were produced. The country realised the importance of thematic maps, and other atlases like irrigation and tourist which were highly appreciated in planning and development. With the production of these atlases, other institutions which have some cartographic potentialities came forward for preparing regional and other atlases. This process of producing regional and thematic atlases is still in progress. About one-third of the Indian states have atlases of one sort or other.

At present the thematic cartography in the country is essentially civilian cartography. The National Atlas and Thematic Mapping Organisation (NATMO) has been premier institutions in this field. However, universities, state government sponsored institutions and private companies have been contributing in thematic cartography as well. But there is a little feedback from these institutions to the NATMO. Mostly, efforts are made in isolation with least interaction. There are duplication of themes and same data base is used by two or more organisations for producing different or similar maps. A mechanism to avoid duplication of efforts has to be worked out. Co-ordination in preparing thematic maps at national level is seriously lacking.

Trend Scenario for the Year 2001 A.D.

Considering the development in thematic cartography in India, it is apparent that this applied field will involve more and more expertise in data handling, data plotting, map designing and meeting requirements of the users. Thematic cartography can no longer be developed in isolation. The advances made in the field of communications, remote sensing, electronic processing of data will certainly influence the thematic mapping technology. Moreover, the users demand for specific themes within limited time frame has to be considered if thematic maps are to be made more meaningful.

The support of technology which needs large infrastructure and large finance is likely to eliminate small time cartographic groups. As a result, by the turn of the century, small thematic cartography groups will not longer be able to stand in the competition which is not a healthy sign at all. Techniques have to be developed with supporting hardware and software facilities which are less expensive so that the smaller institutions can survive. Technology transfer from national institutions and laboratories to smaller institutions and industries has to be an essential part of the thematic cartographic activities in the country.

The data now being used in thematic mapping a voluminous. By A.D. 2001 the conventional and non-conventional data is likely to become more and more extensive, and it will rather become impossible to handle the data manually. For example, the remote sensing data which till recently available was having a resolution of 80 metres. But, now, the resolution has already reached 10 metres. The new data sets will be having more bands as well. On the other hand, such data will be available from a number of sensors which signifies that the capacity of electronic processing of data should be multiple times with large random access memory. Further, the conventional data sets, such as census are becoming more detailed and more refined which increases the volume of data to be handled for preparing thematic maps (Nag, 1984b). Obviously the increasing volume of data can not be ignored.

Two additional aspects are likely to become more important by the turn of the century, they are (a) technology will continue to become sophisticated and more easily available, and (b) expectations of the users including government will become more and more specific. Both the trends are complementary if the progress is made in right direction. The choice of technology will be wideranging which will lead to several options for jump in cartographic technology. The options will not only be in terms of technology, but also in terms of infrastructure. The issues like opening up additional centres, use of remote sensing data and communication system will come up again and again. What are the varieties of outputs which users will appreciate in terms of time, thematic contents, cost and design? Further, thematic mappings is likely to be influenced by the developments and demands in other fields.

Future of the Society in the Beginning of the 21st Century

In 1973 Government of India formed a National Committee on Science

& Technology (NCST) panel on futurology with an aim of starting futuristic studies in India. The panel identified nine areas, viz. (i) energy generation and needs, (ii) housing, (iii) transportation, (iv) communication, (v) urbanology and slum problems, (vi) rural development, (vii) food, (viii) management, and (ix) education with a view to prepare scenario for the year A.D. 2000 presenting likely and desired future prospects and available options in these fields (DST, 1978). Indepth studies of these panels considered issues like social and technological break-through and fast breeder technology, soft and hard technology, micro-processor technology for modularizing various technologies, large scale integrations to possibly place considerable logic and memory on a single chip, landuse, organisation of space, social self management, forest and water policies and other relevant aspects in their totality.

India will soon pass through the 'take-off' stage of economic development. The Indian society is already undergoing rapid transformation with the introduction of technology in every sphere of human activity. Its population will increase but the growth rate will decrease considerable. High mobility will force the rapid changes in the society. Every piece of land will be used for multiple purposes and there would be a significant change in landuse within a day, and shift from the existing socio-economic grouping to socio-technological grouping of population. Land units will have to be identified in terms of technology. The Indian society is likely to transform in the following way :

(a) higher national and per capita incomes,
(b) development of large number of urban centres which will have more uniform distribution pattern,
(c) higher individual, group and inter-settlement connectivity,
(d) transformation of the social setup including increasing number of incomplete families,
(e) ever increasing capacity of individual and society for mass consumption, hence highaer productivity requirements,
(f) frequent breakdown of man-environment relationships,
(g) increased dependency on technology in maintaining the environment,
(h) increase in technological hazards,
(i) extremities in population distribution patterns, and
(j) frequent technological and development shocks.

Obviously to meet the above challenges, maps have to be prepared for carving out development strategies with options and alternatives. Such maps are likely to help in policy oriented researches as well. Moreover, in which way there will be a feedback from policy implications to actual practice will be a matter of greater concern. Thematic maps have to become an integral part of decision-making process and its implementation, so that the mechanism for building the preferred scenario is possible.

Preferred Future for Thematic Cartography

As mentioned earlier, the development of thematic cartography can not be done in isolation. The advancements in other disciplines and technologies will have a direct bearing on the progress of this applied branch of science. More dynamic cartographic techniques will have to be worked out in order to meet the challenges of the 21st century. On the other hand, can valuable and voluminous data be ignored? Remote sensing is becoming a major and homogeneous source of information with improved resolution. Similarly, the conventional sources of data are also becoming more refined. Obviously, in order to handle such a huge data, computers will have to be established in regional and state centres and in data producing agencies. Methods showing multiple use of land according to time and floors have to be worked out. Alongwith time, the third dimension, i.e. height, will become an additional parameter for cartographic considerations. The vertical movement of population is to increase significantly with the construction of single self-sufficient township buildings. Simultaneously, quick data plotting methods are to be introduced in view of preparing realtime thematic maps.

The users should have a variety of choice of getting thematic information for (a) education, where multi-coloured maps with cartographic designing is required; (b) planning, where several overlays for preparing synthetic maps is required; and (c) research, where several experiments are to be conducted. In all types of outputs the question of time is very important. For educational purposes, maps showing themes which are relatively static in nature would be useful. The preparation of these maps will consume time, hence the thematic information should not necessarily be realtime. Moreover, for planning purposes, thematic information must be realtime. Here cartographic design is not important. But, for research, the experiments regarding thematic contents in spatio-temporal framework has to be continued till the researcher is not satisfied with the output. This requires a provision for handling several data sets.

Obviously, the question of technology and proviso for human interaction will become more vital in near future. There can be three strategies : (a) purely manual, (b) purely automated, and (c) a compromise of manual and automated systems. The third strategy is considered to be suitable for developing countries, including India. In such countries, trained or semi-skilled manpower is perhaps available, but the infrastructure is not likely to develop due to financial constrains. Further, following factors may influence the future decision for choosing the technology in the national cartographic institutions in developing countries:

(a) provision for more employment to cartographers,
(b) acquiring superior or intermediate technology for mapping,
(c) availability of external funds with certain conditions, and
(d) nature of users' demands.

It is becoming essential to consider the hardware and software options available or likely to be available in such countries. Computers are improving all over. Their capabilities, storage capacity are assisting operations of various kinds. With the availability of super and fifth generation computers, it is becoming increasingly possible to handle a situation which were considered near-impossible. We are in fourth generation era. It would be relatively easier to simulate a process by taking several defaults and subroutines. This will lead to less human interaction with the computer based systems (Thiagarajan, 1985). It would also be possible to evaluate, synthesize and compare different data sets from diverse sources but for a single theme — an important aspect in thematic mapping.

It has already been experienced that by using appropriate software, throughout can be improved. Lot of researches are being done in order to find out appropriate software for cartographic and remote sensing applications. The software are becoming more and more specialized and certain functions can be done in batch mode. The advancements in the Geographical Information System (GIS) software will certainly assist in thematic cartography. Cumbersome processes of editing, generalization, polygon merging, removal of silver lines, generation of symbols, and other related functions are now being available in standard GIS software packages. What type of software packages be available by the turn of the century, or what is desired ? These questions will have to be thought in details. Perhaps the potentials for developing software are available in some of the developing countries like India which can be explored right now. This will help in promoting a better relationship between thematic cartography institutions and the software industry (Dutt & Nag, 1987).

Preferred Future Scenario

Considering the likely changes in the society, the issues of general importance in developing can be grouped into eight major heads. They are (a) environment, (b) resources, (c) population, (d) agriculture and food production, (e) industrial development, (f) transport, tourism and marketing, (g) administration, and (h) international aspects. Some attempts are being made to map the above themes. From the survey of recent contributions in geography and cartography, it is becoming increasingly evident that serious researches are to be carried out before going for cartographic application or formulating a mapping proposal. Hence, for all major issues, emphasis has to be given on research and methodological aspects as well. Keeping in view of perceived scenario, activities of thematic cartography in 21st century for developing countries in general and India in particular, has been identified (Table 15.1). In order to achieve this, multi-dimensional activities will be required. The component of technology and technology oriented techniques will have their way in every sphere of mapping technology. The idea is not to switch over to a high technology system, but to develop it side by side of the existing conventional system. Perhaps, different phases for all round development can be identified from the experiences of the developed countries or the even the developing countries which have passed through the transitional stage.

Preparation of atlases, such as thematic, regional or planning, is one of the major product of thematic cartography. "Atlases facilitate the cross-correlation of maps not only by making maps of related distributions easier to locate, but also by providing more comparable scales, projections and sets of symbols. . . . The buyers of an atlas

Table 15.1: National Issues and Thematic Mapping

National Issues	Present Mapping Activities	Mapping Activities in 21st Century
1. Environment Hazards	Physical, geomorphological slope, climate, biographical, drainage, wildlife, wet lands,	Research, monitoring and forecasting of Environment, preparation of Environmental Atlases, Himalaya Environmental mapping, Landform mapping, study of regional environmental problem, Pollution of all kinds, floods, droughts, acid rain, heat islands.
2. Resources	Minerals, Water Ocean, Power, Forest	Distribution of minerals, ground-water and surface water, study on the aspects of rural and urban water supply, researches on the changes in the society with conventional and non-conventional sources of power, sea flood, sea resources, sea territorial rights afforestation.
3. Population	Distribution, density, crowding working force, urbanisation, Scheduled Caste and Tribe, age & sex.	Intensive study on the state of national health and diffusion of diseases, Atlas and map of population change, study on the pattern of regrouping of population and socio-technological changes. Socio-economic Atlas of India, population projection.
4. Agricultural & Food Production	Soils, landuse, crops, irrigation plantation, livestock, pressure on land, land capability, land holding, rural settlement.	Research on land resources, Atlas of Rural India, maps and studies of food production systems, agricultural typology, rural development etc.
5. Industrial development	Manufacturing, cement, building materials, chemical, engineering, sugar, handicraft.	Atlas of the Indian Industries and researches on the development of industrial complexes.
6. Transport, Tourism and Marketing	Detailed map of Transport and tourism, railways, roads, post & telecommunication, traffic flow, ports, rural and agricultural marketing, cultural land-scape, history and archaeology.	Historical & Archaeological atlas, research on commodity flow, studies on market infrastructure.
7. Administration and Education	Administrative hierarchy, parliamentary constituency, state maps, physical features.	Current administrative maps and atlases, school atlases, well charts and maps.
8. International Issues	India & Adjacent Countries, Indian Ocean, Arabian Sea, Bay of Bengal.	Atlas of Non-Aligned Movement Countries, Atlas of Indian Ocean Realm, Indian World Atlas, research on developing countries.

purchases not merely a set of maps, but an organized set of maps uniformly portraying comparable data (Monmonier, 1981)". However, it has been forecasted that the world is moving towards a paperless society. The conventional form of maps in loose sheets or atlas may not remain as popular as it is today. Hence, their ratio is likely to decrease as shown in Table 15.2. On the other hand, the non-conventional form of spatial data representation will increase substantially. The non-conventional media will be in the form of (a) digital mode (CCT and CD-ROM), (b) picture mode, (c) annotated imageries, (d) electronic atlases, (e) microfiche atlases, and (f) online map transfer. It is likely that additional methods for data representation will be developed by the turn of the century.

TABLE 15.2: **Estimated Trend of Future Requirements (in %)**

Products	1985	1990	1995	2000	2005	2010
Atlas sheets	90	50	40	30	20	15
Thematic maps	10	40	40	30	25	15
Synthetic maps	—	4	0	20	20	20
Digital mode	—	4	5	10	20	25
Picture mode	—	1	2	4	5	10
Annoted imageries	—	1	2	3	5	5
On line transfer	—	—	1	3	5	10
Total	100	100	100	100	100	100

In general there has been a trend to provide training in all stages of thematic map-making in the national institutions of the developing countries. Cartography has been an essential part of such training programmes, while remote sensing is getting importance due to its wider applicability. With the improvement in technology, it will be necessary to introduce computers and automated printing. The aim should be provide several training courses in order to have fully trained thematic cartographers (Table 15.3)

Action Plan

Once it has been decided that a national cartographic institution dealing with thematic cartography has to move to the 21st century in a desired way, alternatives and options have to be considered along with the infrastructural facilities. It is expected that such institutions will use satellites for following purposes:

(a) remote sensing
(b) transmission of conventional data
(c) interlinkages of different units for data collection
(d) transfer of information from computers to computers or to electronic display units.

TABLE 15.3: **Trend of training needs (% of the scientific staff)**

Type of training	1985	1990	1995	2000	2005	2010
Scribing	20	30	35	35	30	25
Computer familiarization	0.5	30	40	50	60	75
Computer programming	0.2	5	5	5	5	5
Printing	1	5	5	5	5	5
Automated printing	—	1	2	2	2	2
Remote sensing	75	80	85	85	85	85

The satellite should be linked with the mainframe computer which would be processing the data received in a acceptable cartographic form, and the output would be in several forms as mentioned earlier. With the availability of such a super structure, the activities of the thematic mapping institutions will expand in order to include new operations as well. These operations can be continuous online transfer of information collected by conventional or non-conventional means to a crises committee meeting, cabinet meeting or to a research centre. Relay of geographical knowledge in the form of school or university courses through satellite will become possible. However, conventional form of data representation, i.e., thematic maps and atlases should also continue. Maps will be produced much quicker by electronic processing of thematic data, computer finishing of film negatives and automated printing. Four stages can be identified for promoting thematic cartography in developing countries in a desired way (Fig. 15.1).

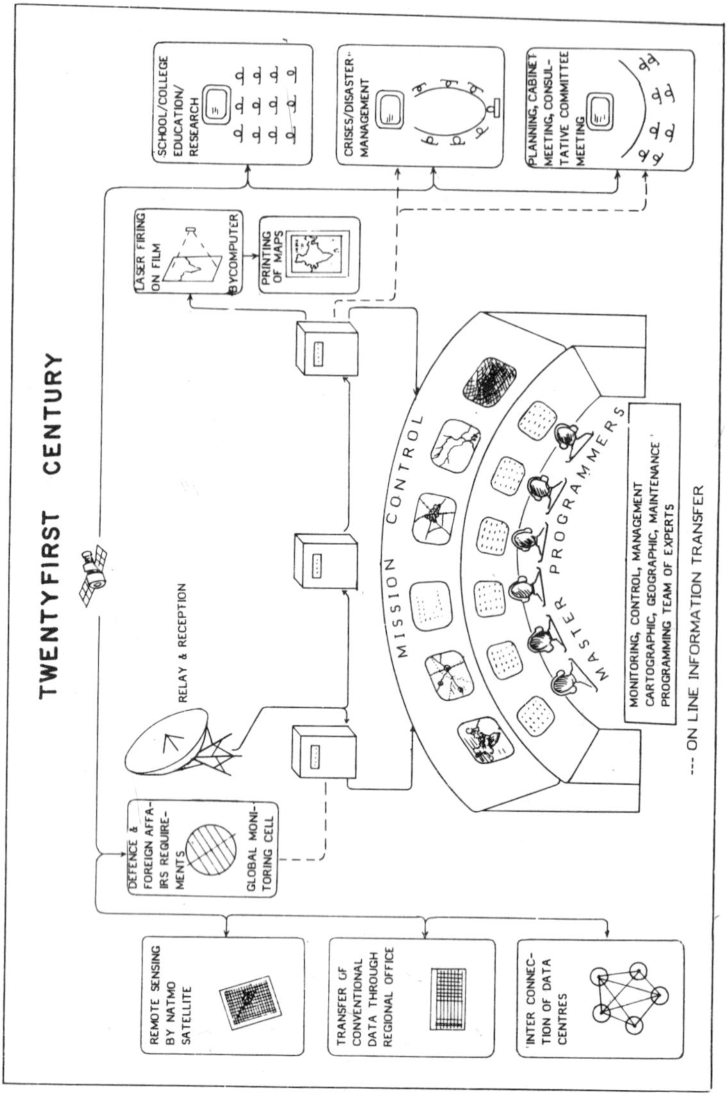

FIG. 15.1: Scenario of Thematic Cartography in 21st Century

Stage One

Probably the most humble beginning in this direction is to develop a micro-processor or personal computer (PC) based system which is capable of handling remote sensing data and digital mapping in a limited way. Inputs can be in the form of CCTs which can be loaded from a tape drive, or by a video digitizer. The processing can be done with the help of V.T. monitor, graphic tablet and by the key board itself. For outpit, dot matrix printer, inkjet plotter or 35 mm film recorder can be used. Since the requirements of thematic mapping are specific, software have either to be procured specifically for the purpose or purchased with the hardware as a package. Obviously, a limited part of thematic map-making can be done on this system.

Stage Two

It is likely that a micro-processor or PC based system will initiate computer culture in the concerned organisation. But, this system has to be developed into a more matured one. At this stage the development can be splitted into (a) preparation of fresh maps, (b) revision of maps, and (c) development of user specific synthetic maps. The inputs can be in the form of statistical data, maps, and point and line details. The proposed system should be able to accept such data. The main sub-systems are digitizer for vector data input, scanner for raster data input, and laser plotter for preparation of film negatives directly. The latter can be used for conventional plate making. However, hard copy for limited users can also be made available with the help of photowrite system.

At this stage, a separate remote sensing system will also be essential. Fortunately, India has advanced in this field and expertise is available in different establishments of the Indian Space Research Organisation. Considering the volume of work, it will be essential to have a dedicated CPU for processing of remote sensing data. The visual interpretation will require standard facilities. The CPU will not only assist in enhancing the image but will be utilized in preparing False Colour Composites (FCCs).

Stage Three

The third stage should concern with integration of digital mapping system with remote sensing system. The integrated system should be able to handle the conventional and non-conventional data. Interfacing between the two system has to be worked out. Information of all sorts should be possible to plot on a cartographic base. The conventional data should thereby be updated by non-conventional information sources. Relevant software programmes will have to be developed. A matured integrated system should be able to develop a geographical information system for a country or a part of it. The data bases and cartographic base in India are ready for this purpose (Nag, 1987).

Stage Four

In the fourth stage, the GIS should be directly linked with the basic source of information through satellite, and conventional data should be received with the help of this satellite through the major and minor centres located throughout the country. The advancement in communication technology will assist in developing a network of centres linked with the next generation of computers. The maser programmers in the control room should be able to handle any situation, apart from printing of maps which will remain the basic activity in thematic cartography, however, laser technology can be used for this purpose. An extra ordinary situation may arise due to crises or disaster management, or due to a querry from central planning or implementation authorities, or to take a decision on an international matter. Further, distance education will become an essential part of the mass education programme in the developing countries where such an integrated system should be able to take part in this activity of vital importance. Obviously, the computer, remote sensing, communication and printing technologies will have to be integrated into one integral system (Fig. 15.1).

Jump in Technology

Above four levels for modernization have been identified. It is not necessary that this process has to be started from stage 1 to stage 4. Perhaps, for India, stage 1 and 2 can be started at a same time. The first one is for initiating computer culture while the other one is to achieve a major jump in technology. The latter will become necessary due to the challenges for specialized jobs likely to come in near future. Further, refinement of techniques and handling of diverse data sets have become an essential part of the thematic mapping technology. Stage 3 can be considered as a "take-off" stage for institutions in developing countries. With the experiences of the earlier stages, the processing of information will be possible. Finally, the feedback from complementary technologies, the inter-linkages will become wide; and a matured system, like the one which has been preferred, is likely to emerge.

It may be possible to jump straight to stage 4, but for this purpose a huge investment at a time has to be done for procurement of hardware, software, infrastructure development and training. It may not be possible for all the developing countries to go for such a huge investment.

Conclusion

The overall activities outlined earlier will make it apparent that persons with specific background can only become a successful thematic cartographer. The core staff in the national institutions will be consisting of persons having knowledge and qualifications in geography and cartography. Systematic training will have to be made compulsory for all new entrants. In India, thematic cartography has reached a stage where specialized culture and infrastructure has already developed with activities and services attached to it. However, in other countries the situation may not be so. The model which India provides is of a human system trying to introduce modernization in all sectors of thematic map-making by A.D. 2001. Such an attempt may appear to be haphazard at the initial stage due to constraints of all sorts which are typical of developing countries. But a move towards a right direction will overcome such attempts by the turn of the century. As the developed countries move towards more and more sophistication, India and other developing countries will find new challenges. Perhaps the developing countries will find themselves in a better position benefiting from the earlier experiences of setback and achievements of developed countries.

REFERENCES

Department of Science & Technology (DST), (1978), *Report of the National Committee of Science & Technology Panel on Futurology*, Government of India, New Delhi.

Dutt, G.K. and P. Nag, (1985), *Life and Work of Professor S.P. Chatterjee*, National Atlas & Thematic Mapping Organisation Occasional Monograph 1, Calcutta.

———, (1987), *National Atlas & Thematic Mapping Organisation in 2001 A.D.*, NATMO Occasional Monograph 2, Calcutta, (in press).

Monmonier, Mark S., (1981), Trends in atlas development, in Guelke, Leonard (ed), *Maps in Modern Geography: Geographical Perspectives on the New Cartography*, Cartographica, 18, No. 2, Monograph 27, pp. 187-213.

Nag, P. (ed), (1984a), *Census Mapping Survey*, IGU Commission on Population Geography, Concept Publishing Company, New Delhi.

———, (1984b), Post-census mapping, in Clarke, John I. (ed), *Geography and Population: Approaches and Applications*, Pergamon Press, Oxford.

———, (1987), A proposed base for geographical information system for India, *International Journal on Geographical Information System*, 1, No. 2.

Seth, S.C., (1975), Seth-Harve method: A scenario building methodology, *Indian Review of Management and Future*, pp. 50-1.

———, (1985), *India in 7000 Days from Now*, John Wiley, New Delhi.

Taylor, D.R.F., (1986), Computer assisted cartography in developing countries, *Proceedings of Auto-Carto London*, Vol. 2, pp. 462-71.

———, (1985), Cartography and international development: Some conclusions, in Taylor, D.R.F. (ed), *Education and Training in Contemporary Cartography*, Chapter 15, John Wiley, pp. 299-313.

APPENDIX I

Glossary and Abbreviations

Analog	–	A form of continuous data recording instead of discrrate fashion.
Bit	–	Binary digit (0 or 1)
Byte	–	8 bits of digital information.
CCTs	–	Computer compatible tapes — magnetic tapes containing digital data of standard 19 cm. wide with 9 track format.
CD — ROM	–	Compact Disk — Read only Memory.
E.S.A.	–	European Space Agency
F.G.C.S.	–	Fifth Generation Computer Systems
FOV	–	Field of View — An angle through which an sensor captures information.
G.C.P.	–	Ground Control Point — Recognizable geographical feature used for controls and geometrical corrections.
Geocoding	–	Geographical referencing of data.
G.I.S.	–	Geographical Information System — A system of capturing, storing, retrieval, integrating, analysing and display of data which can be spatially referred with appropriate software.
G.K.S	–	Graphic Kernal System — A method for allowing software to be used to a variety of different devices.
G.R.E.	–	Ground Resolution Element — Area on the earth surface covered by instantaneous field of view (IFOV).
Hardware	–	The physical components of computer.
IFOV	–	Instantaneous field of view.
Image	–	Object of group of objects recorded by optical, electro — optical, optico — mechanical or electronic means.

ISRO – Indian Space Research Organisation.

LAN – Local Area Network — a interconnecting system joining several nearby processing devices.

LIS – Land Information System — A system for acquiring, processing, storing and disseminating, information about land.

MSS – Multispectral Scanner — A scanning system that receives discrete wave length from earth surface.

NATMO – National Atlas & Thematic Mapping Organisation — A premier organisation in thematic cartography in India, Located in Calcutta.

NNRMS – National Natural Resources Management System — A Planning Commission approved activity for the management and monitoring of the natural resources of the country through remote sensing.

NRSA – National Remote Sensing Agency, located in Hyderabad for receiving remote sensing data from LANDSAT, SPOT, IRS— IA etc. and their supply to users. It has other activities as well.

Optical disc – A computer storage devices using lazer technology to store and retrieve data from such disc coated in light sensitive material.

PC — AT – Personal computer — Advanced Technology.

PC — XT – Personal computer — Extended Technology.

Pixel – One cell on a grid within which data are stored.

Principal Component Analysis – A statistical method for data compression.

RAM – Random Access Memory.

Raster – A regular grid of cells/pixels covering an area and recorded by scanning.

RRSSC – Regional Remote Sensing Service Centre — Five centres located in Dehra Dun, Jodhpur, Kharagpur, Nagpur and Bangalore for processing of the remote sensing data.

RMS – Root Mean Square — used for accuracy calculation.

SAC – Space Applications Centre — A ISRO organisation located in Ahmedabad.

Sensor	–	A device to collect energy, or electromagnetic radiation etc.
Software	–	A computer based programme loaded in computer hardware for specific application, such as data processing, analysis, remote sensing etc.
SPOT	–	System probetoir d'observation del la Terra. A series of high resolution remote sensing satellite launched on February 22nd, 1986.
Vector	–	A value with magnitude and direction which is used for storage of spatial data as pair of coordinates.

APPENDIX II

Guidelines to Fill the Landsat Data Request Form

BOX 1 : List address including PIN/ZIP code. The name of the organization should be clearly identified along with acronym (if any). Listing of Telephone. Telex numbers of Cable/Telegraphic address will help NDC contact you faster in times of need.

BOX 2 : Enter mailing address if it is different from entered at BOX 1.

BOX 3 : Enter your reference.

BOX 4 : Enter NDC account number (if known) to ensure effective response from NDC

BOX 5 : Enter number if several sheets are used for the same order.

BOX 6 : Enter scene identification details

- 6.1 Enter sensor MSS or TM
- 6.2 Enter path/row numbers as per LANDSAT Index map.
- 6.3 Enter sub-scene reference in terms of Latitude and Longitude. (applicable for product code 620 only)
- 6.4 Enter data of pass (if known), alternatively preferred period may be specified (Ex:March 1988 to may 1988), if either date of pass or preferred period is not mentioned, the latest data available supplied.

BOX 7 : Enter product code for desired products as specified in the price list.

BOX 8 : Tick mark the required band for black and white photographic products.

BOX 9 : Enter colour combination of any three bands required for TM False Colour Composite (FCC) product, if nothing is mentioned the standard combination of 2,3 & 4 will be considered.

BOX10 : Enter quantity of products required.

10.1 Quantity of copy of black and white photographic products may be entered here (i.e. depending on the number of bandscenes required)

10.2 If multiple copies of the same product are required the total number of duplicate may be specified here.

BOX11 : Enter the unit price of the product for the 1st copy against it and the discounted price for the duplicate products (if asked for)

BOX12 : Multiply values of BOX 10 and BOX 11 and write the figure against the corresponding line.

BOX13 : Enter maximum cloud cover acceptable. If nothing is mentioned the maximum of 10 per cent cloud cover will be considered.

BOX14 : Enter acceptable data quality. If nothing is mentioned 'GOOD' will be considered.

BOX15 : Enter the applications for which the data is likely to be used. This is requested only for Statistical Purposes.

BOX16 : Enter the user agency to which you belong.

BOX17 : Enter up table (LUT) required, if nothing is mentioned STANDARD LUT will be considered.

BOX18 : Remarks may be written if required.

BOX19 : Put your signature with date.

(*Source* : NRSA, Hyderabad)

N R S A DATA CENTRE
LANDSAT DATA REQUEST FORM

5 SHEET NO. ________

1 NAME ________
ADDRESS ________
____
CITY ________ PIN ________
COUNTRY ________
PHONE ________ TELEX ________
CONTACT PERSON ________

2. **MAIL TO (If different from above)**
NAME ________
ADDRESS ________
____
CITY ________ PIN ________
COUNTRY ________
PHONE ________ TELEX ________

3 YOUR REFERENCE: ________
4 NDC ACCOUNT NO. ________
(If Known) : ________

6. Scene Identification							7	8							9	10		11	12
6.1	6.2	6.3 Sub-Secene				6.4	Product	Bands							Colour	Quantity		Unit	Total
Sensor	Path/	Lat		Lon		Date of	Code								Combi-	10.1	10.2	Cost	Cost
	Row	From	To	From	To	Pass		1	2	3	4	5	6	7	nation	1st Copy	Duplicate	Rs.	Rs.

1. Total this Sheet Rs. ________
2. Total previous Sheet Rs. ________
3. Total (1+2) Rs. ________

15
MAXIMUM CLOUD COVER ACCEPTABLE
10 % ________
25 % ________
ABOVE 25 % ________

14
ACCEPTABLE DATA QUALITY
GOOD ________
BAD ________

for NDC USE ONLY :
REGISTERED AT NDC ON : ________
USER REQUEST No. : ________
TOTAL NUMBER OF PRODUCTS PROCESSED : ________

15
APPLICATIONS
Agricultural/Soil Resources .... Geology/Mineral Exploration ....
Cartography .... Hydrology/Water Resources ....
Environment .... Ocean and Marine Resources ....
Forestry .... Any other (please specify) ....
Geography/Landuse ....

16
USER AGENCY
Central Government Agency .... Industrial ....
State Government Agency .... Individual ....
Academic Institution .... Foreign ....

19
The undersigned has read and accepted the General Terms of supply set out at the back of this sheet.

17 LUT REQUIRED ☐ STANDARD ☐ SCENE SPECIFIC ☐

18 REMARKS (IF ANY) ________

____
(Signature & Date)
Name:

See payment conditions given at the back of this sheet

This form is to be mailed to :
NRSA DATA CENTRE
National Remote Sensing Agency,
Balanagar, HYDERABAD - 500 037, INDIA

TELEPHONE : 262960 or 262572 Ext. 63
TELEX : 0425-6522 GRAMS: REMOSEN

APPENDIX III

Spot Programming Request

General Information

1. For future data acquisition an order programming SPOT data acquisition can be placed on NDC by submitting the Programming Request Form (PR Form). The PR From must be accompanied by the SPOT Data Request Form otherwise the PR Form will not be considered for further processing. The SPOT Data Request Form is to be duly filled in all respects except for the column "Date of Pass" which will be filled by NDC based on the actual date of acquisition. All the columns in the SPOT Data Request Form must be filled to avoid delay in processing for want of clarification. 'NA' may be written wherever not applicable.
2. The PR Form must reach NDC at least 15 days before the beginning of the request period.
3. Similarly, a PR could be cancelled only if intimation is received at NDC 15 days in advance of programming the satellite.
4. The minimum period to be specified for SPOT data acquisition is one month.
5. If the Programming Request cannot be accommodated in the first period, automatically the second or third will be considered.
6. On successful completion of the PR you will be intimated and processing of data shall be taken up to generate the products as specified in the PR form. Once acquired you are bound to buy the products specified by you.
7. Note that the Programming Request will be processed only if sufficient funds are available in your account.

Important: Programming Request Form Received Without Data Request Form will not be Processed.

Spot Processing Levels

RAW FULL SCENE (approx. 60-85 km x 60 km)

-pixel size MLA—20m x 20m, 3000-4250

pixels x 3000 lines
PLA—10m x 10m, 6000-8500 pixels x 6000 lines
-these products are not resampled
-with or without radiometric correction
-no geometric correlation except detector alignment

BULKFULL SCENE (approx. 60-85 km x 60 km)

-pixel size MLA—20m x 20m, 3000-4250 pixels x 3000 lines
PLA—10m x 10m, 6000-8500 pixels x 6000 lines
-these products are not resampled
-radiometric correction
-across-track geometric correction (no GCPs)
-geodetic accuracy — 1.5 km

SYSTEM-CORRECTED GEOCODED (approx. 28 km x 28 km)

-pixel size MLA—12.5m x 12.5m, 2240 pixels x 2240 lines
PLA—6.25m x 6.25m, 4480 pixels x 4480 lines
-radiometric correction
along-track and across-track geometric correction (no GCPs)
geodetic accuracy — 1.5 km

Each HRV produces a 60 km swath. A 120 km can be obtained by positioning the two HRV's appropriately. Users should not that the full 120 km swath can only be obtained by purchasing two (2) scenes.

Guidelines to Fill the Spot Programming Requests Form

BOX 1 : List your address including PIN/ZIP code. The name of the organisation should be clearly identified alongwith acronym (if any). Listing of Telephone, Telex numbers of Cable/Telegraphic address will help NDC contact you faster in times of need.

BOX 2 : Enter mailing address if it is different from that entered at BOX 1.

BOX 3 : Enter your reference.

BOX 4 : Enter NDC account number (if known) to ensure effective response from NDC.

BOX 5 : GEOGRAPHICAL LOCATION OF THE AREA TO BE SURVEYED.

Kindly fill in one PR form for each area to be surveyed. The area to be surveyed may be defined as follows:

1. Survey of India topographic map number, or
2. An area may also be defined by the centres of scenes required using the K, J nodes of the SPOT Reference Grid for areas comprising a maximum of 20 contiguous scenes.

BOX 6 : SPECTRAL MODE

The P + XS mode (simultaneous panchromatic/ multispectral) corresponds to 2 scenes.

BOX 7 : SURVEYING METHOD

— quasi vertical; the viewing angle is comprised between ± 7.5°

— Method with variable viewing angle (± 27°). This method increases the revisit capability in the 26 days cycles. It is recommended for acquisition within short periods of times. This method is mandatory when several acquisitions are required within a cycle or when the acquisition period is critical.

— in the stereoscopic surveying method, the Base/Height ratio is normally close to 1.

BOX 8 : CLOUD COVER

State the maximum cloud cover acceptable scene.

BOX 9 : REQUESTED SURVEYING PERIOD

For standard surveying period (two months or more) indicate the beginning and ending months. For short period, indicate only the month. For critical period mention month and day.

In case you wish one period, indicate also a period of substitution.

When *preparing a PR Form* you must take into account.

— the local average cloud cover during the period:

— the size of area to be surveyed. For large areas several 26 days cycles will be necessary.

— the surveying methods: if the desired viewing angle is

quasi vertical there will be only few surveying opportunities in each 26 day cycles. If, on (±27°), the number of surveying opportunities increases and therefore the acquisition period will shortend, particularly for small areas.

REMEMBER: a stereoscopic acquisition consist of two Surveys.

BOX10 : REQUESTED PRODUCTS FOR THE SCENES ACQUIRED
The Processing level and product code of the desired product may be filled.

BOX11 : Reserved for your observations, if any. It will help if you briefly summarize here the intended purpose of the requested survey.

BOX12 : Put your signature with date.

Guidelines to Fill the Spot Data Request Form

BOX 1 : List your address including PIN/ZIP code. The name of the organisation should be clearly identified alongwith acronym If any). Listing of Telephone, Telex numbers of Cable/Telegraphic address will help NDC contact you faster in times of need.

BOX 2 : Enter mailing address.if it is different from that entered at BOX 1.

BOX 3 : Enter your reference.

BOX 4 : Enter NDC account number (if known) to ensure effective response from NDC.

BOX 5 : Enter sheet number if several sheets are used for the same order.

BOX 6 : Enter scene identification details

6.1 Enter sensor PLA or MLA

6.2 Enter path/row numbers as per SPOT INDEX Map

6.3 Enter date of pass (if known), alternatively preferred period may be specified (Ex, March 1988 to May 1988). If either the date of pass or preferred period is not mentioned, the latest data available will be supplied.

6.4 When Geocoded products are ordered the Survey of India Topographic-Map number is to be indicated. For 1:250,000 and 1:50,000 scale Geocoded products the SOI — Topographic Map number for scale 1:25,000 is to be indicated.

BOX 7 : When Multispectral photographic products are requested the band and the number of copies in each must be entered. The computer compatible tape obviously contain information in all bands. Please make sure that you consider your computer details in selecting tape density.

BOX 8 : The product type and corresponding code, listed in the price schedule, should be determined and entered in the product code column.

BOX 9 : Enter Quantity of products required.

BOX10 : Enter the unit price of the product as specified in the price list.

BOX11 : Multiply values of BOX 9 and BOX 10 and write the figure against the corresponding line.

BOX12 : Enter acceptable data quality. If nothing is mentioned 'GOOD' will be considered.

BOX13 : Enter maximum cloud cover acceptable. If nothing is mentioned the maximum of 10 per cent cloud cover will be considered.

BOX14 : Enter the user agency to which you belong.

BOX15 : Enter the application for which the data is likely to be used. This is requested 'only for Statistical Purposes'.

BOX16 : Remarks may be written here if required.

BOX17 : Put signature with date.

(*Source* : NRSA, Hyderabad)

NRSA DATA CENTRE

SPOT DATA REQUEST FORM

1 NAME ______
ADDRESS ______
______
______
CITY ______ PIN ______
COUNTRY ______
PHONE ______ TELEX ______
CONTACT PERSON ______

2 **MAIL TO (If different from above)**
NAME ______
ADDRESS ______
______
______
CITY ______ PIN ______
COUNTRY ______
PHONE ______ TELEX ______

3 YOUR REFERENCE: ______
4 NDC ACCOUNT NO.
(If Known) : ______

BULK CORRECTED DATA (APPROX. AREA COVERED 60 KM x 60 KM TO 85 KM) 5 SHEET NO.

6.1	6.2		6.3	7 BANDS			8 PRODUCT	9	10	11
SENSOR	PATH	ROW	DATE OF PASS	1	2	3	CODE	QUANTITY	UNIT PRICE	TOTAL PRICE

GEOCODED DATA PRODUCTS (APPROX. AREA 1/4° x 1/4° OR 28 KM x 28 KM)

6.1	6.4 SURVEY OF INDIA	6.2	7 BANDS			8 PRODUCT	9	10	11
SENSOR	TOPOGRAPHIC MAP No.	DATE OF PASS	1	2	3	CODE	QUANTITY	UNIT PRICE	TOTAL PRICE

12 MINIMUM ACCEPTABLE DATA QUALITY
GOOD ______
BAD ______

13 MAXIMUM CLOUD COVER ACCEPTABLE
10 % ______
25 % ______
ABOVE 25 % ______

1. Total this Sheet Rs. ______
2. Total previous Sheet Rs. ______
3. Total (1+2) Rs. ______

14 USER AGENCY
Central Government Agency .... Industrial ....
State Government Agency .... Individual ....
Academic Institution .... Foreign ....

15 APPLICATIONS
Agricultural/Soil Resources .... Geology/Mineral Exploration ....
Cartography .... Hydrology/Water Resources ....
Environment .... Ocean and Marine Resources ....
Forestry .... Any other (please specify) ....
Geography/Landuse ....

17 The undersigned has read and accepted the General Terms of supply set out at the back of this sheet

16 REMARKS (IF ANY) ______

______
(Signature & Date)
Name:

See payment conditions given at the back of this sheet

for NDC USE ONLY :
REGISTERED AT NDC ON : ______
USER REQ[illegible] No. : ______
TOTAL NUMBER OF PRODUCTS PROCESSED : ______

This form is to be mailed to
NRSA DATA CENTRE
National Remote Sensing Agency,
Balanagar, HYDERABAD - 500 037, INDIA

TELEPHONE : 262960 or 262572 Ext. 63
TELEX : 0425-6522 GRAMS: REMOSEN

APPENDIX IV

Guidelines to Fill the IRS Data Request Form

BOX 1 : List your address including PIN/ZIP code. The name of the organisation should be clearly identified along with acronym (if any). Listing of Telephone, Telex numbers or Cable/Telegraphic address will help NDC contact you faster in times of need.

BOX 2 : Enter mailing address if it is different from that centered at BOX 1.

BOX 3 : Enter your reference.

BOX 4 : Enter NDC account number (if known) to ensure effective response from NDC.

BOX 5 : Enter sheet number if several sheets are used for the same order.

BOX 6 : Enter scene identification details

6.1 Enter sensor LISS I or LISS II

6.2 Enter path/row numbers as per IRS Reference Map.

6.3 Enter subscene number (applicable for LISS II scenes)

6.4 Enter date of pass (if known). Alternatively preferred period may be specified in Box 13.

BOX 7 : Enter product Code for desired product as specified in the price list.

BOX 8 : Enter required bands for black and white photographic products.

BOX 9 : Enter colour combination of any three bands required for False Colour Composite (FCC) product.

BOX10 : Enter Quantity of products required.

10.1 Quantity of Ist copy of black and white photographic products may be entered here (i.e. depending on the number of bandscenes required)

10.2 If multiple copies of the same product are required the total number of duplicates may be specified here.

BOX11 : Enter the unit price of the product for the Ist copy against it and the discounted price for the duplicate products (if asked for).

BOX13 : Whether or not date of pass was entered at BOX 6.4 kindly preferred period to enable NDC to select alternative date of pass if required.

BOX14 : Enter maximum cloud cover acceptable.

BOX15 : Enter acceptable data quality.

BOX16 : Enter the applications for which the data is likely to be used.

BOX17 : Enter the user agency to which you belong.

BOX18 : Enter look up table (LUT) required.

BOX19 : Remarks may be written here if required.

BOX20 : Put your signature with date.

General Terms of Supply of Data Products

* ORDERS: No orders will be entertained unless confirmed in writing in any way enabling the purchaser's corporate name to be identified unambiguously. On receipt of the order along with the payment in advance, NDC sends to the purchaser a confirmation containing all the characteristics of the order. No order for a scene normally be amended or cancelled. After the request is registered the purchaser is entitled to no indemnification if the scenes are not obtained or cancelled. After the request is registered the purchaser is entitled to no indemnification if the scenes are not obtained for all or part of the order. NDC reserves the light to refuse any order which it does not consider normally feasible.
* PRICE: The price applicable to each order is the one in effect on the date of confirmation of the order in NDC.NDC publisher a price list of data products at periodic intervals.
* PAYMENTS: All orders must be accompanied by full advance payment for processing to be initiated.
* Payment may be made by demand draft payable to National Remote Agency, Hyderabad or irrevocable letter of credit on State bank of India, Hyderabad — 500 042. andhra Pradesh, India.
* Order and payment may be sent to: NRSA Data Centre, National remote Sensing Agency, Balanagar Hyderabad — 500 037, Andhra Pradesh, India.
* Payment in cash is not accepted.

* For easy operation a standing account can be opened by users by depositing a suitable amount user will be informed of the Account Number. You may add to the balance or obtain a refund of the unused portion at any time. Despatch of data at any time will be limited to the balance amount in your account at that time.
* Payment for all data products shall be in Rupees for Indian users, and in US dollars for overseas users.

Conditions of Sale

* All products are sold for the sole use of purchasers and shall not be loaned, leased or copied without the express permission of and only in accordance with terms and conditions if any agreed with the NRSA Data Centre, National Remote Sensing Agency, Dept. of Space, Govt. of India.
* No conditions of warranties, either express or implied are given or offered for products regenerated by the user at his discretion.
* Data products damaged during transit will be replaced without extra cost. However the quality of the replacement products will not be better than that of the master available with NRSA Data Centre.
* Complaints and Inspection: No complaint relative to the quality and/or quantity of the products delivered will be entertained unless asserted by registered mail within 30 days from the date of despatch. Products may not be returned except with NDC's express prior agreement.
* The purchaser is responsible for any use of the data products purchased from NDC, which has not liability or responsibility for the product's fitness for any particular use. Consequently, the purchaser waives all claims against NDC for any losses, damage, direct or consequential resulting from the use of data products.

  In general all the data products will be despatched by registered insured post/air parcel. Products will be despatched by courier service/speed post at extra cost.
* Unless otherwise specially agreed, all supply of products on the price list are governed by these general terms. No contrary terms or conditions on the purchasers letterhead are binding on NRSA DATA CENTRE.

(*Source*: NRSA, Hyderabad)

IRS-1A ORBITAL CALENDER FOR THE YEAR 1990

PATHS																						
																					1	2
	3	4	5	6	7	8	9	10	11	12	13	14	15	16	17	18	19	20	21	22	23	24
	25	26	27	28	29	30	31	32	33	34	35	36	37	38	39	40	41	42	43	44	45	46
	47	48	94	50	51																	
JAN	1	2	3	4	5	6	7	8	9	10	11	12	13	14	15	16	17	18	19	20	21	22
	23	24	25	26	27	28	29	30	31													
FEB										1	2	3	4	5	6	7	8	9	10	11	12	13
	14	15	16	17	18	19	20	21	22	23	24	25	26	27	28							
MAR																1	2	3	4	5	6	7
	8	9	10	11	12	13	14	15	16	17	18	19	20	21	22	23	24	25	26	27	28	29
	30	31																				
APR			1	2	3	4	5	6	7	8	9	10	11	12	13	14	15	16	17	18	19	20
	21	22	23	24	25	26	27	28	29	30												
MAY											1	2	3	4	5	6	7	8	9	10	11	12
	13	14	15	16	17	18	19	20	21	22	23	24	25	26	27	28	29	30	31			
JUN																				1	2	3
	4	5	6	7	8	9	10	11	12	13	14	15	16	17	18	19	20	21	22	23	24	25
	26	27	28	29	30																	

(Contd.)

(Contd.)

JUL						1	2	3	4	5	6	7	8	9	10	11	12	13	14	15	16	17
	18	19	20	21	22	23	24	25	26	27	28	29	30	31								
AUG															1	2	3	4	5	6	7	8
	9	10	11	12	13	14	15	16	17	18	19	20	21	22	23	24	25	26	27	28	29	30
	31																					
SEP		1	2	3	4	5	6	7	8	9	10	11	12	13	14	15	16	17	18	19	20	21
	22	23	24	25	26	27	28	29	30													
OCT										1	2	3	4	5	6	7	8	9	10	11	12	13
	14	15	16	17	18	19	20	21	22	23	24	25	26	27	28	29	30	31				
NOV																			1	2	3	4
	5	6	7	8	9	10	11	12	13	14	15	16	17	18	19	20	21	22	23	24	25	26
	27	28	29	30																		
DEC					1	2	3	4	5	6	7	8	9	10	11	12	13	14	15	16	17	18
	19	20	21	22	23	24	25	26	27	28	29	30	31									

Source: NNRMS Bulletin

IRS-1A ORBITAL CALENDER FOR THE YEAR 1991

PATHS																					
							1	2	3	4	5	6	7	8	9	10	11	12	13	14	15
16	17	18	19	20	21	22	23	24	25	26	27	28	29	30	31	32	33	34	35	36	37
38	39	40	41	42	43	44	45	46	47	48	94	50	51								
JAN 1	2	3	4	5	6	7	8	9	10	11	12	13	14	15	16	17	18	19	20	21	22
23	24	25	26	27	28	29	30	31													
FEB									1	2	3	4	5	6	7	8	9	10	11	12	13
14	15	16	17	18	19	20	21	22	23	24	25	26	27	28							
MAR															1	2	3	4	5	6	7
8	9	10	11	12	13	14	15	16	17	18	19	20	21	22	23	24	25	26	27	28	29
30	31																				
APR		1	2	3	4	5	6	7	8	9	10	11	12	13	14	15	16	17	18	19	20
21	22	23	24	25	26	27	28	29	30												
MAY										1	2	3	4	5	6	7	8	9	10	11	12
13	14	15	16	17	18	19	20	21	22	23	24	25	26	27	28	29	30	31			
JUN																			1	2	3
4	5	6	7	8	9	10	11	12	13	14	15	16	17	18	19	20	21	22	23	24	25
26	27	28	29	30																	

(Contd.)

(Contd.)

JUL						1	2	3	4	5	6	7	8	9	10	11	12	13	14	15	16	17
AUG	18	19	20	21	22	23	24	25	26	27	28	29	30	31	1	2	3	4	5	6	7	8
	9	10	11	12	13	14	15	16	17	18	19	20	21	22	23	24	25	26	27	28	29	30
SEP	31	1	2	3	4	5	6	7	8	9	10	11	12	13	14	15	16	17	18	19	20	21
OCT	22	23	24	25	26	27	28	29	30	1	2	3	4	5	6	7	8	9	10	11	12	13
NOV	14	15	16	17	18	19	20	21	22	23	24	25	26	27	28	29	30	31	1	2	3	4
	5	6	7	8	9	10	11	12	13	14	15	16	17	18	19	20	21	22	23	24	25	26
DEC	27	28	29	30	1	2	3	4	5	6	7	8	9	10	11	12	13	14	15	16	17	18
	19	20	21	22	23	24	25	26	27	28	29	30	31									

Source: NNRMS Bulletin

NRSA DATA CENTRE

IRS DATA REQUEST FORM

[1] NAME ____
ADDRESS ____
____
____
CITY ____ PIN ____
COUNTRY ____
PHONE ____ TELEX ____
CONTACT PERSON ____

[2] **MAIL TO (If different from above)**
NAME ____
ADDRESS ____
____
____
CITY ____ PIN ____
COUNTRY ____
PHONE ____ TELEX ____

[3] YOUR REFERENCE: ____
[4] NDC ACCOUNT NO. ____
(If Known) : ____

for NDC USE ONLY :
REGISTERED AT NDC ON : ____
USER REQUEST No. : ____
TOTAL NUMBER OF PRODUCTS PROCESSED : ____

This form is to be mailed to
NRSA DATA CENTRE
National Remote Sensing Agency,
Balanagar, HYDERABAD - 500 037, INDIA

TELEPHONE : 262960 or 262572 Ext. 63
TELEX : 0425-6522 GRAMS: REMOSEN

[5] SHEET NUMBER:

[6] Scene Identification				[7] Product Code	[8] Bands				[9] Colour Combination	[10] Quantity		[11] Unit Cost	[12] Total Cost
6.1 Sensor	6.2 Path/row	6.3 Sub-scene	6.4 Date of Pass		1	2	3	4		10.1 Ist copy	10.2 Duplicate	Rs.	Rs.
1													
1													
1													
1													
1													
1													
1													
1													
1													

1. Total this Sheet Rs. ____
2. Total previous Sheet Rs. ____
3. Total (1+2) Rs. ____

[13] PREFERRED PERIOD
Jan - Mar ...... Years:
Apr - Jun ...... From ...... To ......
Jul - Sep ...... Specific coverage:
Oct - Dec ...... From ...... To ......

[14] MAXIMUM CLOUD COVER ACCEPTABLE
10 % ____
25 % ____
ABOVE 25 % ____

[15] ACCEPTABLE DATA QUALITY
GOOD ____
BAD ____

[16] APPLICATIONS
Agricultural/Soil Resources .... Geology/Mineral Exploration ....
Cartography .... Hydrology/Water Resources ....
Environment .... Ocean and Marine Resources ....
Forestry .... Any other (please specify) ....
Geography/Landuse ....

[17] USER AGENCY
Central Government Agency .... Industrial ....
State Government Agency .... Individual ....
Academic Institution .... Foreign ....

[20] The undersigned has read and accepted the General Terms of supply set out at the back of this sheet.

[18] LUT REQUIRED ☐ STANDARD ☐ SCENE SPECIFIC ☐

[19] REMARKS (IF ANY) ____

____
(Signature & Date)
Name:

See payment conditions given at the back of this sheet

APPENDIX V

Announcement of Satellite Data Products Prices for 1990-91

Because of significant cost escalation for maintenance and consumables, the upward price revision has become inevitable. The price revision has been done taking into account the promotional aspects and the limited number of users activity involved in remote sensing. The revised prices will become effective on September 1, 1990 and will be valid till March 31, 1991. For orders placed before August 31, 1990, the existing prices are applicable.

The revised prices for 1990-91 are as follows:

IRS Data Products Price List For 1990 — 91

Standard and Geocoded Products

Photographic products are available on Cubic convolution resampling and SOM/POL map projections for both LISS I and LISS II.

Product Code		Product Type	Nominal Scale / Format		Unit Price	
SOM	POL		LISS-I	LISS-II	In Rs. for Indian users	In US$ for foreign users
BLACK & WHITE PHOTO PRODUCTS PER SCENE PER BAND						
0201	0301	Film negative 70mm	1:2,892,000	1:1,446,000	1600	160
0211	0311	Film positive 70mm	1:2,892,000	1:1,446,000	1600	160
0202	0302	Film negative 240mm	1:1,000,000	1:500,000	500	50
0212	0312	Film positive 240mm	1:1,000,000	1:500,000	500	50
0222	0322	Paper print 240mm	1:1,000,000	1:500,000	500	50
0223	0323	Paper print 480mm	1:500,000	1:250,000	700	70
0224	0324	Paper print 960mm	1:250,000	1:125,000	900	90
—	—	Geocoded paper print 5 times enlargement	Not offered	1:50,000	700	70
FALSE COLOUR COMPOSITE PHOTO PRODUCTS PER SCENE						
0242	0342	Film positive 240mm	1:1,000,000	1:500,000	1600	160
0252	0352	Paper print 240mm	1:1,000,000	1,500,000	1600	160
0253	0353	Paper print 480mm	1:500,000	1:250,000	1800	180
0254	0354	Paper print 960mm	1:250,000	1:125,000	2700	270
—	—	Geocoded paper print 5 times enlargement	Not offered	1:50,000	2000	200

(Contd.)

(Contd.)

Product Code		Product Type	Nominal Scale / Format		Unit Price	
SOM	POL		LISS-I	LISS-II	In Rs. for Indian users	In US$ for foreign users
COMPUTER COMPATIBLE TAPE PER SCENE						
0266	0366	Standard correction nearest neighbour resampling	1600/6250 BPI	1600/6250 BPI	3000	300
0466	0566	Standard correction cubic convolution resampling	1600/6250 BPI	1600/6250 BPI	3000	300
—	—	Geocoded standard correction Polyconic projection	Not offered	1600/6250 BPI	2300	230
CARTRIDGE PRODUCT PER SCENE						
BIL	BSQ					
016C	017C	Standard correction nearest neighbour resampling	40 MB	40 MB	3000	300
016C	017C	Standard correction cubic convolution resampling	40 MB	40 MB	3000	300
026C	027C	Geocoded standard correction Polyconic projection	No offered	40 MB	2300	230
FLOPPY PRODUCT PER SEGMENT						
2GA	02GA	LISS-I Standard, 1k x 1k single band, 1.2 MB			*500	50
02GA	02GA	LISS-II Standard, 512 x 512, 4 bands, 1.2 MB			*500	50
02GA	02GA	LISS-II Geocoded, 512 x 512, 4 bands, 1.2 MB			*500	50

(Contd.)

(Contd.)

Product Code		Product Type		Nominal Scale / Format		Unit Price	
SOM	POL			LISS-I	LISS-II	In Rs. for Indian users	In US$ for foreign users
PRECISION PRODUCTS							
BLACK & WHITE PHOTO PRODUCTS PER SCENE PER BAND							
0602	0702	Film negative		1:1,000,000	1:500,000	800	80
0612	0712	Film positive		1:1,000,000	1:500,000	800	80
0622	0722	Paper print		1:1,000,000	1:500,000	800	80
0623	0723	Paper print	2 times enlargement	1:500,000	1:250,000	1100	110
0624	0724	Paper print	4 times enlargement	1:250,000	1:125,000	1400	140
FALSE COLOUR COMPOSITE PHOTO PRODUCTS PER SCENE							
0642	0742	Film positive		1:1,000,000	1:500,000	2400	240
0652	0752	Paper print		1:1,000,000	1:500,000	2400	240
0635	0753	Paper print	2 times enlargement	1:500,000	1:250,000	2700	270
0654	0754	Paper print	4 times enlargement	1:250,000	1:250,000	4100	410
COMPUTER COMPATIBLE TAPE PER SCENE							
0666	0766	Precision corrected cubic convolution		1600 BPI / 9 track	1600 BPI /	4500	450
0866	0966	Precision corrected nearest neighbour		1600 BPI / 9 track	1600 BPI / 9 track	4500	450

(Contd.)

(Contd.)

Product Code SOM	POL	Product Type		Nominal Scale / Format LISS-I	LISS-II	Unit Price In Rs. for Indian users	In US$ for foreign users
SPECIAL PRODUCTS							
1. FULL SCENE							
1.1 BLACK AND WHITE PHOTO PRODUCTS PER SCENE PER BAND							
—	×302	Film negative		1:1,000,000	1:500,000	800	80
—	×312	Film positive		1:1,000,000	1:500,000	800	80
—	×322	Paper print		1:1,000,000	1:500,000	800	80
—	×323	Paper print	2 times enlargement	1:500,000	1:250,000	1100	110
—	×324	Paper print	4 times enlargement	1:250,000	1:250,000	1400	140
1.2 FALSE COLOUR COMPOSITE PHOTO PRODUCTS PER SCENE							
—	×342	Film positive		1:1,000,000	1:500,000	2400	240
—	×352	Paper print		1:1,000,000	1:500,000	2400	240
—	×353	Paper print	2 times enlargement	1:500,000	1:250,000	2700	270
—	×354	Paper print	4 times enlargement	1:250,000	1:250,000	4100	410
2. SUB — IMAGE							
2.1 SIZE 512 x 512							
—	1×323	B&W	Paper print	—---	1:50,000	1100	110
—	1×353	FCC	Paper print	—---	1:50,000	2700	270
2.2 SIZE 1024 x 1024							
—	2×324	B&W	Paper print	—---	1:50,000	1400	140
—	2×354	FCC	Paper print	—---	1:50,000	4100	410

APPENDIX VI

Floppy Products of Satellite Data

In most of the P.C. based image processing systems floppy disk is widely used as input medium rather than magnetic tapes. Due to availability of low cost P.C. based image processing systems, satellite data on small media like floppydisk will be handy for analysis purpose. Keeping this in view we now introduce new satellite data products, namely, "FLOPPY PRODUCTS".

Most of the P.Cs marketed are IBM compatible, operating on MS-DOS and Xenix operating systems and they support BASIC and FORTRAN. Therefore, the format is so designed that floppy can be read by BASIC/FORTRAN under the above operating system.

Structure of Layout of Floppy Product

The layout of the floppy product is illustrated in the diagram given below:

MASTER DESCRIPTION RECORD	MASTER BLOCK
HEADER DESCRIPTION RECORD HEADER RECORD	HEADER BLOCK
IMAGE DESCRIPTION RECORD IMAGE RECORDS	IMAGE BLOCK

The data is arranged in three major blocks within the file. They are designated as MASTER, HEADER and IMAGE blocks. Each block has one description record and zero or more logical records. Each logical record will be 512 bytes long and consists of physical records of 256 bytes at drive level. Whereas the description record provides information on the available fields on the records.

All the fields in the Master and Header blocks are in ASCII format and each field within a record is terminated by a field separator. The field separator is the combination of two ASCII characters — "carriage return" and "line feed". This allow all the dump of descriptive information to be got directly on the terminal or printer.

Within the image block, description record is in ASCII and Imagery data is in binary, with no record separators.

Types of Floppy Products

The floppy products of various satellites that are proposed to be supplied on the standard 5.25 inch floppies of 1.2 MB are listed in Table A 5.1.

In all the products listed in the Table A 5.1, a CCT scene is subdivided into number of segments to accommodate the data in the limited capacity of the floppy. One floppy has the data on one segment only. The area for which data is required may be specified in terms of corner coordinates in latitude and longitude. Depending upon the number of segments that are being covered under the specified area the costing of the data products is done at the rates mentioned in Table A 5.1 per floppy.

The segment coding scheme for the floppy product is explained below by a few example;

1. **Product Size of 1K×1K**: Suppose consider record length of the TM data is 3900 x 2848 bytes and segment of 1K x 1K (1024 x 1024 bytes) has to be extracted then the labelling of the segments is done as shown in figure below:

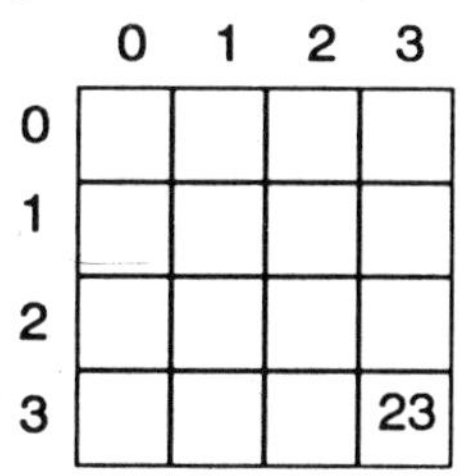

As we note that since the record size (3900 x 2848) is not an exact multiple of 1K, the following can be observed:

1. Row numbers: 0, 1, 2
2. Column number: 0, 1, 2, 3
3. An example segment 23 is shown

4. All segments in column 3 consist of 828 pixels and 1K lines except segment 23
5. All segments in row 2 consist of 800 lines and 1K pixels except segment 23
6. Segment 23 consists of 828 pixels and 800 lines
7. All other segments are 1K x 1K

II. **Product size of 512 x 512**: The segment coding for 512 x 512 bytes for TM data of record length 3900 x 2848 bytes is shown in figure below:

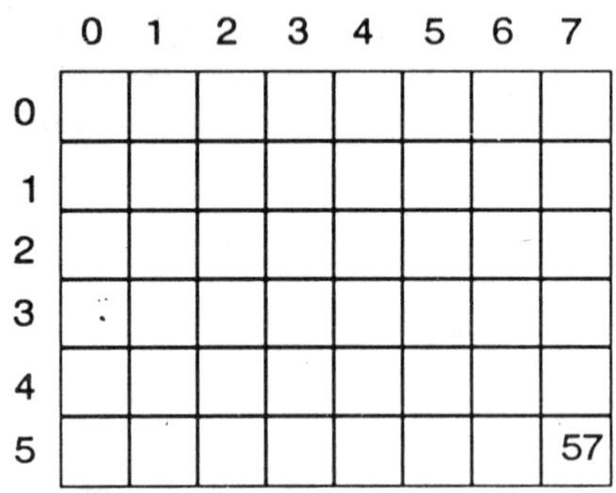

Following may be observed:

1. Row numbers: 0, 1, 2....5
2. Column number: 0, 1, 2....7
3. An example segment 57 is shown
4. All segments in column 7 consist of 316 pixels and 512 lines except segment 57
5. All segments in row 5 consist of 288 lines and 512 pixels except segment 57
6. Segment 57 consists of 316 pixels and 288 lines
7. All other segments are 512 x 512

The record lengths of CCT data of various satellites with Grid specifications is summarised in the table given in Table A 5.2.

The order for the Floppy Products may be placed in the enclosed 'Floppy form" clearly specifying the area of interest in terms of latitude and longitude.

It may be noted that:

1. the floppy products are supplied only with "Bulk Correction".
2. the floppy products can be supplied for a single band based on the user's requirement.
3. If the data requested shifts from the standard segment(s) it becomes necessary to order for more than one segment covering the area of interest. In such cases the cost will be charged for all the segments supplied.

TABLE A5.1:
Floppy Products Price List

Product Code	Satellite	Sensor	Product size: *Segment size	Product size: No. of bands	Floppy Capacity	Cost per Product Rs.	**Discounted price Rs.
02GA	***IRS-1A	LISS-I	1K X K	1	1.2 MB	500	300
02GA		LISS-IIA	512 X 512	4	1.2 MB	500	300
02GA		LISS-IIB	512 X 512	4	1.2 MB	500	300
02GA	SPOT-1	PLA	1K X 1K	1	1.2 MB	500	300
02GA		MLA	1K X 1K	1	1.2 MB	500	300
02GA		MLA	512 X 512	3	1.2 MB	500	300
02GA	Landsat-5	TM	1K X 1K	1	1.2 MB	650	450
02GA		TM	512 X 512	4	1.2 MB	650	450

* A scene is subdivided into number of segments to accommodate the data in the limited size of floppy capacity.

** Discounted prices are applicable for the floppies ordered from the existing Floppy library.

*** Discount of 25% is allowed on all IRS products for the first year, i.e., 1988-89.

Note

1. Prices are effective from November 1, 1988 and are valid upto March 31, 1989
2. Prices are subject to change without notice.
3. Prices are inclusive of mailing costs

TABLE A5.2:
Floppy Products
Record Length and Grid Specification

SAT	Sensor	Product Size	Record Length	Labelling of Segments along Pixel	Labelling of Segments along Scanline	Total Number of Segments
IRS	LISS-I	1K X 1K	2350	0 TO 2	0 TO 2	3 x 3 = 9
IRS	LISS-II	512 X 512	2350	0 TO 4	0 TO 4	5 X 5 = 25
SPOT	PLA	1K X 1K	6480	0 TO 6	0 TO 5	7 X 6 = 42
SPOT	PLA	1K X 1K	9000	0 TO 8	0 TO 5	9 X 6 = 54
SPOT	MLA	1K X 1K	3600	0 TO 3	0 TO 2	4 X 3 = 12
SPOT	MLA	512 X 512	3600	0 TO 6	0 TO 5	7 X 6 = 42
SPOT	MLA	1K X 1K	3960	0 TO 3	0 TO 2	4 X 3 = 12
SPOT	MLA	512 X 512	3960	0 TO 7	0 TO 5	8 X 6 = 48
SPOT	MLA	1K X 1K	4500	0 TO 4	0 TO 2	5 X 3 = 15
SPOT	MLA	512 X 512	4500	0 TO 8	0 TO 5	9 X 6 = 54
LAND-SAT	TM	1K X 1K	3900	0 TO 3	0 TO 2	4 X 3 = 12
LAND-SAT	TM	512 X 512	3900	0 TO 7	0 TO 5	8 X 6 = 48

Guidelines to fill the Floppy Request Form

Note: The serial numbers below correspond to the number in the order form.

1. List your address including PIN/ZIP code. The name of the organisation should be clearly identified, alongwith its acronym (if any). Listing of telephone/telex numbers or cable/telegraphic address will help NDC to contact you faster in times of need.
2. If the mailing address is different from your address it should be clearly specified so.
3. In case you have had previous business with NRSA Data Centre, please mention your customer code number also in all your correspondence to ensure effective response from NDC.
4. Sheet Number to be filled-in if several sheets are used for the same order.
5. Scene Identification.

5.1 Write sensor name: MLA or PLA in case of SPOT, TM in case of LANDSAT, LISS-IIAI, LISS-IIA2, LISS-IIB1 or LISS-B2 in case of IRS.

5.2 Write required area in terms of path and row
5.3 Write the band(s) for which data is required
5.4 Write the latitudinal extents
5.5 Write the longitudinal extents
5.6 Write the date of acquisition only if specifically required, alternatively the period of time in which data should have been acquired may be listed.

6. Write the code of the required product as given in the price list.
7. Indicate in this column the number of floppies.
8. Prices as per the price list.

All other blocks to be filled-in appropriately to facilitate the final selection or to select an alternate date by NDC personnel.

To order a Floppy from the Existing Library

Fill in the order form as explained above except for the following;

5.4 Write the segment number instead of latitudinal extents, and
5.5 Write the Floppy-Id instead of longitudinal extents.

(*Source*: NNRMS Bulletin)

NRSA DATA CENTRE

FLOPPY REQUEST FORM

1 NAME ____________
ADDRESS ____________
____
____
CITY ________ PIN ________
COUNTRY ____________
PHONE ________ TELEX ________
CONTACT PERSON ____________

2 **MAIL TO (If different from above)**
NAME ____________
ADDRESS ____________
____
____
CITY ________ PIN ________
COUNTRY ____________
PHONE ________ TELEX ________

3 YOUR REFERENCE: ____________
NDC ACCOUNT NO. ____________
(If Known) : ____________

for NDC USE ONLY :
REGISTERED AT NDC ON : ____________
USER REQUEST No. : ____________
TOTAL NUMBER OF PRODUCTS PROCESSED : ____________

This form is to be mailed to :
NRSA DATA CENTRE
National Remote Sensing Agency,
Balanagar, HYDERABAD - 500 037, INDIA

TELEPHONE : 262960 or 262572 Ext. 63
TELEX : 425-2076 GRAMS: REMOSEN

4 SHEET NUMBER												
5 SCENE IDENTIFICATION									6	7	8	9
5.1 SENSOR	5.2 PATH-ROW	5.3 BAND(S)	5.4 LATITUDE SEGMENT NO		5.5 LONGITUDE FLOPPY-ID		5.6 DATE OF PASS		PRODUCT CODE	QUANTITY	UNIT COST	TOTAL COST
			FROM	TO	FROM	TO					Rs.	Rs.

1. Total this Sheet Rs. ________
2. Total previous Sheet Rs. ________
3. Total (1+2) Rs. ________

PREFERRED PERIOD

Jan - Mar ☐	Years:		
Apr - Jun ☐	From	To	
Jul - Sep ☐	Specific coverage:		
Oct - Dec ☐	From	To	

MINIMUM CLOUD COVER ACCEPTABLE
10 % ☐
30 % ☐
50 % ☐
70 % ☐
OTHERS (SPECIFY)

APPLICATIONS

AGRICULTURAL/SOIL RESOURCES	☐	GEOLOGY/MINERAL EXPLORATION	☐
CARTOGRAPHY	☐	HYDROLOGY/WATER RESOURCES	☐
ENVIRONMENT	☐	OCEAN AND MARINE RESOURCES	☐
FORESTRY	☐	ANY OTHER (PLEASE SPECIFY)	☐
GEOGRAPHY/LANDUSE	☐		

USER AGENCY

CENTRAL GOVERNMENT AGENCY	☐	INDUSTRIAL	☐
STATE GOVERNMENT AGENCY	☐	INDIVIDUAL	☐
ACADEMIC INSTITUTION	☐	FOREIGN	☐
....			

REMARKS (IF ANY) ____________

(Signature & Date) ____________

See payment conditions given at the back of this sheet

Name:

APPENDIX VII

Global Thematic Maps on Environment

1. Relief of the surface of the earth, 1:39 million, 1985
2. General bathymetric map of the oceans, 1:15 million, 1984
3. Ocean currents and drainage basins, 1:20 million, 1978
4. Average significant wave height, to order, 1988
5. Bedrock geology of the world, 1:23 million, 1985
6. Tectonic map of the world, 1:15 and 1:45 million, 1984
7. Global tectonic and volcanic activity over the last million years, 1:27 million, 1979
8. Significant earthquakes 1900-1979, 1:30 million, 1981
9. Average annual precipitation, to order, 1988
10. Global surface temperature, by day, to order, 1988
11. Average annual short-wave radiation, 1:30 million, 1969
12. Annual evapotranspiration, 1:20 and 1:50 million, 1977
13. Major ecosystems complexes, 1:30 million, 1983
14. Annual average vegetation index, to order, 1988
15. Ocean chlorophyl, to order, 1988
16. Biogeographic provinces, 1:39 million, 1975
17. Types of soils, 1:25 million, 1983
18. Desertification, 1:25 million, 1977
19. Soil degradation, 1:15 million, in preparation for 1990
20. Arctic and Antarctic sea ice, to order, 1988

Additionally, two general maps, population density (alternatively world by night, showing distribution of settlements) and land use in preparation, 1989

Source: IGU/ICA Committee Report on the Thematic Maps of the World.

APPENDIX VIII

List of Global Digital Datasets on Environmental Thematic

\+ * 1. Global land elevation and terrain
2. World elevation and bathmetry
3. World elevation, bathmetry and sea ice
* 4. Coast, river and lakes
5. Earthquakes
6. Geomagnetic annual means
7. Main field geomagnetic data
\+ 8. Global aeromagnetic survey 1953-72
\+ 9. Global aeromagnetic survey 1981-83
10. Significant earthquakes
11. Earth's magnetic field
12. Surface air temperature variations
\+ * 13. Ozone distribution
\+ * 14. Temperature anomalies
\+ * 15. Surface temperature
\+ * 16. Rainfall anomalies
17. Evapotranspiration and precipitation
18. CO_2 from fossil fuel burning
19. Dust veil index
20. Surface air temperature 1851-1984
21. Annual and seasonal temperature anomalies
22. Annual temperature variations 1861-1984
23. seasonal and annual radiosonde records
24. Ozone deviations 1958-1985
\+ * 25. Albedo (four seasons)
* 26. Soils and soil characteristics (FAO/UNESCO)
27. Soils — US Department and nitrogen
\+ * 29. Vegetation index
30. Natural vegetation
31. Tropical forest resource assessment
32. Net primary production

33. World ecosystems complexes
* 34. Cultivation intensity
35. Primary and secondary land cover
36. Terrestrial hydrography

* Data sets already in GRID
\+ From satellite imagery

Source: IGU/ICA Committee Report on the Thematic Maps of the World.

APPENDIX IX

List of Scientific Journals Related to Thematic Cartography and Remote Sensing (1991)

- American Cartographer
- Annals of the Association of American Geographers
- Annals of the National Association of Geographers of India
- Boletin della Associazione Italiana de Cartografia
- Bibliographia Cartographica
- Boletin de Cartografia
- Canadian Cartography
- Canadian Journal of Remote Sensing
- Canadian Surveying and Mapping Journal
- Cartactual
- Cartographica
- Cartography
- Cartographic Journal
- Computer Vision, Graphics and Image Processing
- Geographical Review of India
- Geographical Review
- Geographical Analysis
- Geografiska Annalar
- Geojournal
- Geoprocessing
- Globus
- Geodezia es Kartografia
- Imago Mundi
- International Journal on Geographical Information Systems
- International Journal on Remote Sensing
- International Yearbook on Remote Sensing
- International Yearbook on Cartography
- ITC Journal
- ISPRS Journal of Photogrammetry and Remote Sensing Interface
- Journal of Geography

— Journal of Applied Photographic Engineering.
— Kartographische Nachrichten
— Map
— National Geographical Journal of India
— NNRMS Bulletin
— Photogrammetrica
— Photogrammetric Record
— Photogrammetric Engineering and Remote Sensing
— Photonirvachak
— Polski Przeglad Kartograficzny
— Remote Sensing of Environment.

APPENDIX X

Indian Remote Sensing Satellite IRS-1C/1D

IRS 1C/1D projects are planned as continuation projects of IRS 1A/1B with improved resolutions. These projects have been approved by the Government of India in May 1989 with the following objectives:

1. To design, develop, launch and operate state-of-the-art three axis body stabilised satellite for providing continued space based remote sensing services to the user community with enhanced resolution and capability compared to IRS-1A/1B.
2. Further, develop new areas of user applications to take full advantage of the enhanced resolution and capability of IRS-1C/1D spacecrafts.

   IRS-1C is planned to be launched in 1993 and 1D in 1995. Like its predecessors, IRS-1C is a three axis stabilised spacecraft to support new payloads. The tentative modification of the payload under considerations are:

   (a) a multi-spectoral camera (called LISS-3) with about 23m. resolution working in 3 bands (green, red, infrared) and having 140 km. swath. Additionally a middle infrared camera of 69m resolution and a swath of 140 km. is planned.

   (b) a panchromatic camera with a resolution of better than 10m and a swath of 70 km. which is steerable by ± 26° across track.

   (c) A piggy back payload Wide Field Sensor (called WIFS) operating in two bands (red and infrared) providing a resolution of about 180m with a swath of 770 km is also planned as a complementary payload to the multi-spectoral camera given above.

Mission

A polar Sun-Synchronous orbit of 817 kms has been chosen with a descending node and equatorial cross over at 10:30 a.m. The multi-sectoral camera (LISS-3) will provide a repetitivity of 24 days and the panchromatic camera can provide revisit capability of 5 days. The mission/payload specifications are given in the annexure.

Spacecraft bus

The spacecraft bus is configured around IRS-1A bus with mission specific modifications in power system, TTC, AOCS and data handling systems. IRS-1C main frame will be built around IRS-1A bus to use the heritage and get as much reliability as possible. Two-band carriers are introduced to transmit the data from the payloads. An onboard tape recorder is added in order to provide capability to record 24 minutes of either panchromatic or LISS-3 data for payback in the subsequent orbit over Indian ground stations.

Ground Segment

The ground segment consist of Mission Control, Payload Data Reception and Data Processing Centres. The mission specific changes in the ground segment are in terms of the type and quantum of data products to be generated and the associated data processing systems as well as the modified data reception system. The mission control will be similar to IRS-1A. The new data product software will have to be developed to suit the data requirements.

Annexure

Mission Specifications

Orbit

Orbit	–	817 km (Sun Synchronous)
Node	–	Descending node
Equatorial cross over time	–	10.30 a.m.
Payloads	–	(1) LISS-3 (2) PAN (3) WIFS

Payload Specifications

1. *LISS-3 Camera*

		Bands	Resolutions
Bands	–	(i) 0.52-0.59μm	23m
		(ii) 0.62-0.68μm	23m
		(iii) 0.77-0.86μm	23m
		(iv) 1.55-1.7 μm	69m
Swath	–	140 km	
Repetitivity	–	24 days	
SNR	–	128	
Quantisation	–	7 bits	
Calibration	–	Relative calibration with CED	

2. *Panchromatic Camera*

Band	–	0.5-0.75 μm
Geometric resolution	–	Better than 10 metres
Swath	–	70 km
Steerability	–	$\pm 26^\circ$
Quantisation	–	6 bits
SNR	–	64
Calibration	–	Relative calibration with CED
Revisit	–	5 days

3. *WIFS Camera*

Bands	–	(i) 0.62-0.68 μm
	–	(ii) 0.77-0.86 μm
Resolutions	–	180 metres
Swath	–	770 km
Quantisation	–	7 bits
Repetitivity	–	5 days

4. *Spacecraft platform*

Pointing accuracy	–	Roll & pitch 0.15° Yaw $\pm 0.2^\circ$
Stability of pointing	–	3 × 10E 4 deg/sec.
Attitude determination	–	$.08^\circ$ with conical sensors $.01^\circ$ with star sensors
Orbit determination	–	500 metres
Data Transmission	–	X-band
TTC	–	S-band

Source: NNRMS Bulletin

BIBLIOGRAPHY

Berry, J.K., (1987), Fundamental operations in computer assisted map analysis, *International Journal of Geographical Information System*, Vol. 1(2), pp.119-36.

Cracknell, A. and Ladson Hayes (eds), (1990), *Remote Sensing Yearbook, 1990*, Taylor & Francis, London.

Cracknell, A.P. and *et al.* (eds), (1989), *Computer Programs for Image Processing of Remote Sensing Data*, University of Dundee, Dundee.

Cracknell, A.P. and L.W.D. Hayes, (1990), *Introduction to Remote Sensing*, Taylor & Francis, London.

Curran, Paul J., (1985), *Principles of Remote Sensing*, English Language Book Society/Longman.

Dale, Peter F. and John D. Mc Laughtin, (1988), *Land Information Management*, Clarendon Press, London.

Elachi, C., (1987), Information to the *Physics and Techniques of Remote Sensing*, John Wiley & Sons, New York.

Eason, Ken, (1989), *Information Technology and Organisational Change*, Taylor & Francis, London.

Estes, John E. and Leslie W. Senger (eds), (1974), *Remote Sensing: Techniques for Environmental Analysis*, Hamilton Publishing Company, Santa Barbara.

Goodchild, Michael F. and Sucharita Gopal (eds), (1984), *The Accuracy of Spatial Databasis*.

Hemdon, Col. Robert E., (1961), Aerospace cartography, *Surveying and Mapping*, Vol. 21, pp.31-43.

Jensen, John R., (1986), *Introductory Digital Image Processing – A Remote Sensing Perspective*, Prentice – Hall, New Jerse.

Molenaar, Martian, (1989), Towards a geographic information theory, *ITC Journal*, Vol. I, pp.5-11.

Mathur, P.M., (1977), *Computer Processing of Remotely – Sensed Images*, John Wiley & Sons, London.

Mackay, I. Ross, (1962), Some cartographic problems in the field of special (thematic) maps, *Canadian Cartography*, Vol. I, pp.42-7.

Mounsey, Helen and Roger Tomlinson (eds), (1988), *Building Databases for Global Science*, Taylor & Francis, London.

Muller, Jan-Peter (ed), (1988), *Digital Image Processing in Remote Sensing*, Taylor & Francis, London.

Nag, P. (ed), (1984), *Census Mapping Survey*, Concept Publishing Company, New Delhi.

Nag, P., (1987), A proposed base for the development of GIS for India, *International Journal on Geographical Information System*, Vol. 1, No. 2, pp.

Pannakock, A.J., (1968), Problems involved in producing the National Atlas of Netherlands, *Cartographic Journal*, Vol. 5, pp.8-15.

Parry, R.B. and C.R. Perkins, (1987), *World Mapping Today*, Butterworths, London.

Peuquet, Donna J. and Duane F. Marble (eds), (1990), *Introductory Readings in Geographical Information System*, Taylor & Francis, London.

Rhind, D.W., I.S. Evans and M. Visvalingam, (1980), Making a National Atlas of Population by computer, *Cartographic Journal*, Vol. 17, No. 1, pp.3-11.

Rhind, David and Helen Mounsey, (1990), *Understanding Geographic Information System*, Taylor & Francis, London.

Raper, Jonathan (ed), (1989), *Three Dimensional Applications in GIS*, Taylor & Francis, London.

Star, Jeffrey and John Estes, (1990), *Geographical Information System*, An Introduction, Prentice Hall, New Jersey.

Sabins, F.F., (1978), *Remote Sensing: Principles and Interpretation*, W.H. Freeman & G. San Francisco.

Salichtchev, K.A., (1979), Periodical and Serial publications on cartography, cartographica: *The Canadian Cartographer*, Vol. 16, No. 2, pp.109-32.

Taylor, D.R.F. (ed), (1985), *Education and Training in Contemporary Cartography*, John Wiley & Sons Ltd.

Taylor, D.R.F., (1979), Development atlases using computer techniques, *Cartographica*, Monograph 23, pp.67-78.

Tomlinson Roger F., (1989), Canadian GIS experience, *Canadian Institute of Surveying and Mapping Journal*, Vol. 43, No. 3, pp.227-32.

Tomlinson, R.F., (1987), Current and potential uses of geographical information system The American Experience, *International*

Journal on Geographical Information Systems, Vol. 1(3), pp.203-18.

Tobler, W.R., (1965), Automation in the preparation of thematic maps, *Cartographic Journal*, Vol. 2, pp.32-8.

Vanghan, R.A. and R.P. Kirby (eds), (1989), *Geographical Information Systems and Remote Sensing for Local Resource Planning*, University of Dundee, Dundee.

Ward, Candra E., (1988), 1990 Censuses to increase use of automation, *Asian and Pacific Population Forum*, Vol. 2, Nos. 3-4.

Zubrow, E.; K. Allen and S. Green (eds), (1990), *Interpreting Space: GIS in Archaeology and Anthropology*, Taylor & Francis, London.

INDEX